Strategy and Tactics

NS310

Strategy and Tactics

Mark Mill and Alex Mamikonian
Editors

U.S. Naval Academy
Division of Professional Development
Department of Seamanship and Navigation

Naval Institute Press
Annapolis, Maryland

Naval Institute Press
291 Wood Road
Annapolis, MD 21402

Library of Congress Cataloging-in-Publication Data

NS 310 : strategy and tactics / U.S. Naval Academy, Division of
Professional Development, Department of Seamanship and
Navigation.
    p. cm.
  ISBN 1-59114-523-6 (alk. paper)
  1. Naval strategy—Textbooks. 2. Naval tactics—Textbooks.
I. Title: Strategy and tactics. II. Title: NS310. III. United States
Naval Academy. Dept. of Seamanship and Navigation.
  V163.N75 2005
  359.4—dc22
                                                    2005016742

Printed in the United States of America on acid-free paper ⬯

12 11 10 09 08 07 06 05   9 8 7 6 5 4 3 2
First printing

All photos are courtesy of the *U.S. Naval Institute Photo Archive.*

Contents

Introduction: Professional Knowledge

As Midshipmen, you are members of the American military profession. In his landmark study, *The Soldier and the State*, Samuel Huntington, defines attributes of a profession as responsibility, expertise, and corporateness.[1] That is, members of a profession have a sense of responsibility to society (often made explicit by an oath or obligation, such as the Hippocratic Oath for doctors or the Commissioning Oath for naval officers). Professionals are recognized experts in their field. Finally, they are members of a group distinct from society, with a distinct code of conduct or ethics. For instance, lawyers are members of the bar in their home state; military officers are made distinct by their uniforms, regulations, and customs.

All professionals have an obligation to continually enhance their expertise in their chosen field. For doctors, this means the study of new discoveries and techniques in medicine; for lawyers the continual review of recent laws and legal decisions. As members of the military profession, we are obligated to continually enhance our professional expertise. Journals such as *Proceedings, Marine Corps Gazette*, and *Joint Forces Quarterly*, professional societies, and continual education provide plenty of opportunities for the Sailor, Midshipman, or officer to increase his knowledge and preparation for war.

---

[1] Samuel P. Huntington, *The Soldier and the State: The Theory and Politics of Civil-Military Relations* (Cambridge: Harvard University Press, 1967), p. 8.

Yet, here at the Naval Academy, Midshipmen often meet the term professional knowledge with derision. Because plebes are required to learn "pro" knowledge, and anything endured during the plebe year has a negative connotation, students often arrive in NS 310 dreading the professional knowledge that is a vital part of the course. Yet, as we see above, we are obligated by our choice to enter the military to become experts in the profession of arms. Professional study is the path to achieve that end, and it is a constant for all members of the profession, from the seaman studying for his petty officer's exam to the admiral reading the latest article in the *Naval War College Review*. Plebe year introduces us to the terminology of the military: the difference between a cruiser and a frigate, the meaning of the chain of command. We will not repeat plebe year in this course; instead we will assume that you know—and hold you accountable for knowing—the information studied then. (If you do not know it, reviewing that information with *your* plebes is an excellent way to review!) In this course, we will begin to apply that knowledge, taking the first steps towards making the decisions required of us as leaders and commanders in the Naval Service.

Course Policy

I. CLASSROOM POLICY

The following policies apply for my classes.

- Be accountable! If you have any problems with any assignments, other classes, other commitments, etc., let me know and something can be worked out.
- Be prepared for class. Come to class having read the assignment and prepared to discuss the topics with me and your classmates.
- Be on time for class! If you are late, you will be marked tardy.
- No sleeping in class. We are all professionals and will treat each other as such.
- Drinks are allowed, BUT must be in a can or *covered* container.
- Working on other assignments during class will not be tolerated. If you have difficulty managing your time, see me and we will work something out.
- Always maintain a sharp military appearance.
- Classes are a military obligation. If you have to miss a class, notify me. Call or e-mail me *prior* to class.
- Changing class periods is not authorized, *no exceptions*.
- Should I be late for or absent from class, wait 10 minutes and send a representative to Room 316 for further instructions.

The Honor Concept always applies...

II. CLASS TEXTS & MATERIALS

Required texts:
The NS310 Strategy and Tactics Textbook

Additional course information is available at:
Blackboard for NS310 Fall 2005

III. HOMEWORK & QUIZZES

Students are responsible for all assignments (both reading and written), as well as all material covered in class despite an absence during that particular lesson. If I announce on Monday that an assignment is due Wednesday, it is due Wednesday at the commencement of class, even if the student missed Monday's class.

All homework, reading assignments and quizzes are assigned in accordance with the syllabus, unless otherwise directed by me or a substitute instructor.

IV. PRESENTATIONS

Midshipmen will be required to present a project on a Current Event/ Naval Accident or Mishap.
Guidelines for the presentations are on Blackboard.

V. GRADING POLICIES

Graded Event	6 Week	12 Week	16 Week	Final Grade
Exam 1	80%	40%	35%	25%
Exam 2	–	40%	35%	25%
Presentations	–	-	20%	15%
Instructor Input Homework Quizzes Class Participation	20%	20%	10%	10%
Final Exam	–	–	–	25%
Total	100%	100%	100%	100%

Strategy and Tactics

1

Carrier Strike Group

In the *CNO Guidance for 2003* Admiral Vernon Clark stipulated that the terms "Carrier Battle Groups" and "Amphibious Readiness Groups" would no longer be the standards terms and that they would be replaced by Carrier Strike Groups and Expeditionary Strike Groups, respectively, by March 2003. The goal being to find ways to effectively produce naval capability in a more efficient manner.

Under this initiative, Cruiser-Destroyer and Carrier Groups are designated as Carrier Strike Groups (CSG) and aligned directly under the numbered fleet commanders. This realignment gives key operational leaders authority and direct access to the people needed to more effectively accomplish the Navy's mission. Formerly, Carrier Group (CARGRU) and Cruiser-Destroyer Group (CRUDESGRU) staffs were under the administrative authority of the air and surface type commanders (TYCOM). With this new initiative, authority and control will come from the numbered fleet commanders who are responsible for the training and certification of the entire Strike Group. The organizational structure to support the Carrier Strike Groups focuses more on placing Strike Group commanders under the authority of the certifying officer, or the numbered fleet commander.

Under this concept, the warfare distinction of either the air-side or the surface-side is removed and is unified as Carrier Strike Groups.

The Carrier Strike Group (CSGS) provides the full range of capabilities that were present in carrier battle groups. It remains the joint task force commander's premier power projection option. However, because surface combatants will be needed for Expeditionary Strike Groups and Surface Action Groups, the number of ships escorting the carrier would be reduced.

In the new concept, the CSGS would deploy with three or four surface combatants, all Aegis ships. With the introduction of an improved E-2C Hawkeye aircraft and CEC, these ships would provide the group with sufficient defense against the most likely air, surface and subsurface threats.

In larger scale conflict or higher threat scenarios, combining multiple CSGSs with SAGs and ESGs would provide the level of combat capability, power projection and force protection required. This consolidated group is known as the expeditionary strike force (ESF).

It is important to note that there really is no real definition of a strike group. Strike groups are formed and disestablished on an as needed basis, and one may be different from another. However, they all are comprised of similar types of ships. Typically a carrier strike group might have:

- a carrier—The carrier provides a wide range of options to the U.S. government from simply showing the flag to attacks on airborne, afloat and ashore targets. Because carriers operate in international waters, its aircraft do not need to secure landing rights on foreign soil. These ships also engage in sustained operations in support of other forces.
- two guided missile cruisers—multi-mission surface combatants. Equipped with Tomahawks for long-range strike capability.
- a guided missile destroyer—multi-mission suface combatant, used primarily for anti-air warfare (AAW)
- a destroyer—primarily for anti-submarine warfare (ASW)
- a frigate—primarily for anti-submarine warfare (ASW)
- two attack submarines—in a direct support role seeking out and destroying hostile surface ships and submarines
- a combined ammunition, oiler, and supply ship—provides logistic support enabling the Navy's forward presence: on station, ready to respond

The CSGS could be employed in a variety of roles, all of which would involve the gaining and maintenance of sea control:

- Protection of economic and/or military shipping.
- Protection of a Marine amphibious force while enroute to, and upon arrival in, an amphibious objective area.
- Establishing a naval presence in support of national interests.

Aside from the renumbering of the Strike Groups, the actual change will directly affect only the administrative chain of command for the 14 CSG staffs. The ships and hardware remain administratively under their current platform TYCOM.

Strike Group commanders remain accountable to the numbered fleet commanders for integrated and sustainment training, and to the TYCOM for materiel readiness and unit (basic) level training of Strike Group units. The CSGs have been renumbered with respect to Navy tradition, with even numbers on the east coast and odd numbers on the west. To preserve their current recognized role as training groups, CSG 1 and 4 have retained their numbers.

Carrier Groups (CCG) and Cruiser-Destroyer Groups (CCDG) will be renamed commander, Carrier Strike Groups (CCSG).

Aircraft Carriers—CV, CVN

Updated: 2 March 2005

Description: Aircraft carriers provide a wide range of possible response for the National Command Authority.

The Carrier Mission

- To provide a credible, sustainable, independent **forward presence** and conventional deterrence in peacetime,
- To operate as the cornerstone of joint/allied maritime expeditionary forces in times of crisis, and
- To operate and support aircraft attacks on enemies, protect friendly forces and engage in sustained independent operations in war.

Features: The aircraft carrier continues to be the centerpiece of the forces necessary for *forward presence*. Whenever there has been a crisis, the first question has been: "Where are the carriers?" Carriers support and operate aircraft that engage in attacks on airborne, afloat, and ashore targets that threaten free use of the sea; and engage in sustained operations in support of other forces.

Aircraft carriers are deployed worldwide in support of U.S. interests and commitments. They can respond to global crises in ways ranging from peacetime presence to full-scale war. Together with their on-board air wings, the carriers have vital roles across the full spectrum of conflict.

John C. Stennis (CVN 74)

The *Nimitz*-class carriers, eight operational and two under construction, are the largest warships in the world. USS *Nimitz* (CVN 68) was the first to undergo its initial refueling during a 33-month Refueling Complex Overhaul at Newport News Shipbuilding in Newport News, Va., in 1998. The next generation of carrier, CVN 21, the hull number will be CVN 78, is programmed to start construction in 2007 and is slated to be placed in commission in 2014 to replace USS *Enterprise* (CVN 65 which will be over its 50-year mark. CVN 79 is programmed to begin construction in

2012 and to be placed in commission in 2018, replacing *USS John F. Kennedy* (CV 67) in her 50th year.

Point of Contact:
Public Affairs Office
Naval Sea Systems Command
Washington, DC 20362

General Characteristics, *Nimitz* Class

Builder: Newport News Shipbuilding Co., Newport News, Va.
Power Plant: Two nuclear reactors, four shafts
Length, overall: 1,092 feet (332.85 meters)
Flight Deck Width: 252 feet (76.8 meters)
Beam: 134 feet (40.84 meters)
Displacement: Approx. 97,000 tons (87,996.9 metric tons) full load
Speed: 30+ knots (34.5+ miles per hour)
Aircraft: 85
Cost: about $4.5 billion each

Cruisers—CG

Updated: 3 December 2004
Description: Large combat vessel with multiple target response capability.
Features: Modern U.S. Navy guided missile cruisers perform primarily in a Battle Force role. These ships are multi-mission [Air Warfare (AW), Undersea Warfare (USW), and Surface Warfare (SUW)] surface combatants capable of supporting carrier battle groups, amphibious forces, or of operating independently and as flagships of surface action groups. Cruisers are equipped with *Tomahawk* cruise missiles giving them additional long range strike mission capability.
Background: Technological advances in the *Standard* Missile coupled with the AEGIS combat system in the *Ticonderoga* class cruisers have increased the AAW capability of surface combatants to pinpoint accuracy from wave-top to zenith. The addition of *Tomahawk* in the CG-47 has vastly complicated unit target planning for any potential enemy and returned an offensive strike role to the surface forces that seemed to have been lost to air power at Pearl Harbor.

The lead ship of the class, USS *Ticonderoga* (CG 47) was decommissioned on 30 September 2004.

Point of Contact:
Department of the Navy (OP-03PA)
Washington, D.C. 20350-2000

General Characteristics, *Ticonderoga* Class

Builders:
Ingalls Shipbuilding: CG 47-50, CG 52-57, 59, 62, 65-66, 68-69, 71-73
Bath Iron Works: CG 51, 58, 60-61, 63-64, 67, 70.
Power Plant: Four General Electric LM 2500 gas turbine engines; 2 shafts, 80,000 shaft horsepower total.
SPY-1 Radar and Combat System Integrator: Lockheed Martin.
Length: 567 feet
Beam: 55 feet
Displacement: 9,600 tons (9,754.06 metric tons) full load
Speed: 30 plus knots
Aircraft: Two SH-2 *Seasprite* (LAMPS) in CG 47-48; Two SH-60 *Sea Hawk* (LAMPS III)
Cost: About $1 billion each

Destroyers—DD, DDG

Updated: 23 March 2005
Description: These fast warships provide multi-mission offensive and defensive capabilities, and can operate independently or as part of carrier battle groups, surface action groups, amphibious ready groups, and underway replenishment groups.

Features: Destroyers and guided missile destroyers operate in support of carrier battle groups, surface action groups, amphibious groups and replenishment groups. Destroyers primarily perform anti-submarine warfare duty while guided missile destroyers are multi-mission [Anti-Air Warfare (AAW), Anti-Submarine Warfare (ASW), and Anti-Surface Warfare (ASUW)] surface combatants. The addition of the Mk-41 Vertical Launch System or *Tomahawk* Armored Box Launchers (ABLs) to many Spruance-class destroyers has greatly expanded the role of the destroyer in strike warfare.

Background: Technological advances have improved the capability of modern destroyers culminating in the *Arleigh Burke* (DDG 51) class. Named for the Navy's most famous destroyer squadron combat commander and three-time Chief of Naval Operations, the *Arleigh Burke* was commissioned July 4, 1991, and was the most powerful surface combatant ever put to sea. Like the larger *Ticonderoga* class cruisers, DDG 51's combat systems center around the Aegis combat system and the *SPY-lD,* multi-function phased array radar. The combination of Aegis, the Vertical Launching System, an advanced anti-submarine warfare system, advanced anti-aircraft missiles and *Tomahawk*, the *Burke* class continues the revolution at sea.

The DDG 51 class incorporates all-steel construction. In 1975, the cruiser USS *Belknap* (CG 26) collided with USS *John F. Kennedy* (CV 67). *Belknap* suffered severe damage and casualties because of her aluminum superstructure. On the basis of that event, the decision was made that all future surface combatants would return to a steel superstructure. And, like most modern U.S. surface combatants, DDG 51 utilizes gas turbine propulsion. These ships replaced the older *Charles F. Adams* and *Farragut*-class guided missile destroyers.

The *Spruance*-class destroyers, the first large U.S. Navy warships to employ gas turbine engines as their main propulsion system, are undergoing extensive modernizing. The upgrade program includes addition of vertical launchers for advanced missiles on 24 ships of this class, in addition to an advanced ASW system and upgrading of its helicopter capability. *Spruance*-class destroyers are expected to remain a major part of the Navy's surface combatant force into the 21st century.

Point of Contact:
Public Affairs Office
Naval Sea Systems Command (OOD)
Washington, DC 20362

General Characteristics, *Arleigh Burke* class

Builders: Bath Iron Works, Ingalls Shipbuilding
Power Plant: Four General Electric LM 2500-30 gas turbines; two shafts, 100,000 total shaft horsepower.
SPY-1 Radar and Combat System Integrator: Lockheed Martin
Length:
  Flights I and II (DDG 51-78): 505 feet (153.92 meters)

Flight IIA (DDG 79-98): 509½ feet (155.29 meters)

Beam: 59 feet (18 meters)

Displacement:

Hulls 51 through 71: 8,315 tons (8,448.04 metric tons) full load

Hulls 72 through 78: 8,400 tons (8,534.4 metric tons) full load

Hulls 79 and on: 9,200 tons (9,347.2 metric tons) full load

Speed: in excess of 30 knots

Aircraft: None. LAMPS III electronics installed on landing deck for coordinated DDG 51/helo ASW operations

2

Expeditionary Strike Group

The Expeditionary Strike Group—sometimes called an Expeditionary Strike Force—is a revamped amphibious ready group with the ability to disperse strike capabilities across a greater range of the force, increasing the striking power in the amphibious ready group.

ESGs would enable the fleets to cover more parts of the world effectively, providing highly mobile, self-sustaining forces that are able to undertake missions across the entire spectrum of operations. The ESG concept could almost double the number of independent operational groups the Navy can deploy in the future, from 19 to 38.

The ESG concept allows the Navy to field 12 Expeditionary Strike Groups and 12 Carrier Battle Groups, in addition to surface action groups.

This concept is based on an earlier one, introduced in the early 1990s, called a Naval Expeditionary Task Force that was also based on the idea that naval forces could be grouped in varying combinations.

The expeditionary strike group—made up of amphibious ships, cruisers, destroyers and submarines—is a departure from the typical carrier battle group/amphibious ready group structure. An expeditionary strike group could include amphibious ships, a destroyer, cruiser, frigate, attack submarine and a P-3C *Orion* land-based aircraft. The new mix, which deploys in place of the amphibious ready group, allows Navy and Marine Corps forces to launch Marines and landing craft as warships and submarines strike inland targets with missiles and shells. Currently, each amphibious ready group is made up of an amphibious assault ship, a dock landing ship and an amphibious transport dock. Cruisers and destroyers deploy with carrier battle groups.

Originally, Navy officials indicated that ESGs would not replace ARGs as the Navy does not have enough ships for 12 CSGs and 12 ESGs (13 upon the arrival of LHD-8). However, by early 2004, several reports in the civilian press had indicated that the Navy was looking to have all of the LHA/D's at the center of an ESG. This intention appears to have changed by late April 2004, after an unidentified Navy official outlined a plan to reduce the number of ESGs from 12 to 8 so that additional funds could be made available for other programs. The ESG cut would be gradual, and would first appear in the FY 2006 budget.

During the conflict in Afghanistan the Navy experimented with adding capabilities to the traditional three-ship ARGs, two of which were operating together in the North Arabian Sea region at the height of the conflict. The Navy attached to its ARGs near Afghanistan Aegis-equipped surface warships, a USS *Spruance* (DD-963)-class destroyer; a submarine for intelligence, surveillance and reconnaissance (ISR); and the links to the P-3 AIP aircraft.

The first Expeditionary Strike Groups began deploying for the first time in 2003. The first deployments are to be a pilot program and will be studied before any changes occur fleetwide.

An Expeditionary Strike Group centered around the USS *Saipan* and the 22nd MEU was to be the first ESG to deploy, followed by the *Peleliu* ESG from the West Coast. Iraqi Freedom altered this timetable resulting the first ESG being centered around the *Essex* and followed by the *Peleliu*. In the June 16, 2003 edition of the *Navy Times* it was reported that the *Saipan* would not be the first Atlantic Fleet ESG as the *Saipan* would be in the yard for maintenance following her role in Iraqi Freedom. Instead, the USS *Wasp* would deploy.

The *Saipan* would have deployed in August 2003 with its usual compliment of LPDs and LSDs plus one other vessel of unknown type but will borrow three other ships from the Enterprise Battle Group which is to deploy roughly around the same time. The ships assigned to the *Saipan* ESG will be LPD 15 *Ponce*, LSD 44 *Gunston Hall*, CG 58 *Philippine Sea*, DDG 66 *Gonzalez*, FFG 47 *Nicholas*, and SSN 755 *Miami*. It will be commanded by a commodore which is generally what would command an Amphibious Ready Group.

With the delay in the *Saipan's* deployment the USS *Peleliu* was designated to become the first ship to deploy as the center of an ESG. The

Peleliu ESG will be commanded by a rear admiral in an experiment to determine which of the two options works out best. The *Peleliu* deployed on August 22, 2003 (ahead of its original November 2003 deployment date) with six other vessels that were to have been detached from a carrier, and the *Peleliu* was to deploy at roughly the same time as a Carrier Battle Group. Originally the *Peleliu* ESG was to include LPD 8 *Dubuque*, LSD 42 *Germantown,* CG 73 *Port Royal*, DDG 73 *Decatur*, FFG 33 *Jarrett* and SSN 754 *Topeka*. The *Topeka* was later replaced by the USS Greenville and the *Dubuque* was replaced by the *Ogden*.

The Inter-Deployment Training Cycle had been altered in that an ESGEX will be added. It is unclear if this would take place instead of a JTFEX as details are still developing.

Amphibious Assault Ships LHA/LHD/LHA(R)

Updated: 30 December 2003

Description: The largest of all amphibious warfare ships; resembles a small aircraft carrier; capable of Vertical/Short Take Off and Landing (V/STOL), Short Take Off Vertical Landing (STOVL), Vertical Take Off and Landing (VTOL) tiltrotor and Rotary Wing (RW) aircraft operations; contains a welldeck to support use of Landing Craft Air Cushion (LCAC) and other watercraft.

Features: Modern U.S. Navy Amphibious Assault Ships project power and maintain presence by serving as the cornerstone of the Amphibious Readiness Group (ARG)/Expeditionary Strike Group (ESG). A key element of the *Seapower 21* pillars of Sea Strike and Sea Basing, these ships transport and land elements of the Marine Expeditionary Brigade (MEB) with a combination of aircraft and landing craft.

The *Tarawa*-class LHA provides the Marine Corps with a superb means of ship-to-shore movement by helicopter in addition to movement by landing craft. Three LHAs—which have extensive storage capacity and can accommodate both LCUs and LCACs—were unusually active during *Operations Desert Shield/Storm*. Since that time, LHAs (and, later, LHDs) have been participants in major humanitarian-assistance, occupation, and combat operations in which the United States has been involved. Such operations have included providing support to NATO forces engaged in keeping the peace in Bosnia, taking part in rescue operations in the offshore waters of African countries ravaged by civil war, and in Kosovo in 1999, and participating in *Operation Enduring Freedom* in the Arabian Sea and the Gulf of Oman in 2001 and 2002. Also, during 2000,

USS *Peleliu* (LHA 5)

USS *Essex* (LHD 2) swapped forward-deployed naval force assignments with USS *Belleau Wood* (LHA 3) as the "big-deck" amphibious ship in Sasebo, Japan. USS *Iwo Jima* (LHD 7) was commissioned in June 2001, and had her first deployment in 2003.

In April 2002 a construction contract was awarded for LHD 8 (*Makin Island*) with contract delivery to the Navy scheduled no later than 31 July 2007. In 2003, the majority of the amphibious assault ships participated in *Operation Iraqi Freedom*, conducting concurrent Well Deck and Flight Deck operations as an integral part of the multi-national forces operations.

In 2003, USS *Peleliu* (LHA 5) deployed as centerpiece of an Expeditionary Strike Group (ESG), introducing a new concept of operations, replacing the Amphibious Ready Groups (ARGs). With delivery of LHD 7, the Navy and Marine Corps has a flexible force of ships—LHAs/LHDs, LPDs, and LSD 41/49s—that can provide 12 fully capable Expeditionary Strike Group forces to fulfill anticipated Marine Corps Lift and forward-presence requirements. The amphibious capability of the fleet will be improved with construction of LHD 8 and the replacement of the *Austin*-class LPDs by *San Antonio*-class LPDs.

Background: Amphibious warships are designed to support the Marine Corps tenets of Operational Maneuver From the Sea (OMFTS) and Ship to Objective Maneuver (STOM). They must be able to sail in harm's way and provide a rapid buildup of combat power ashore in the face of opposition. Because of their inherent capabilities, these ships have been and will continue to be called upon to also support humanitarian and other contingency missions on short notice. The United States maintains the largest and most capable amphibious force in the world. The WASP-class are currently the largest amphibious ships in the world. The lead ship, USS *Wasp* (LHD 1) was commissioned in July 1989 in Norfolk, Va. LHA Replacement or LHA(R) is the next step in the incremental development of the "Big Deck Amphib." She is being designed to accommodate the Marine Corps' future Air Combat Element (ACE) including F-35B Joint Strike Fighter (JSF) and MV-22 *Osprey*, provide additional vehicle and cargo stowage capacities and enable a broader, more flexible Command and Control capability.

Program Status: All LHAs are in-service; LHDs 1-7 are in-service, LHD 8 is under construction and expected to deliver in July 2007. LHAR program is in the early stages. The lead LHAR is planned for delivery to the Fleet in 2013.

Point of Contact:
Public Affairs Office
Naval Sea Systems Command
Washington, DC 20362

General Characteristics, LHA(R) Class

Builder: TBD (currently undergoing functional design)
Power Plant: Two marine gas turbines, two shafts, 70,000 total brake horsepower

Length: 921 feet (280.7 meters)

Beam: 116 feet (35.4 meters)

Displacement: Approx. 50,100 long tons (50,905 metric tons) full load

Speed: 20+ knots

Aircraft, Depending on mission: a mix of: F-35B Joint Strike Fighters (JSF) STOVL aircraft; MV-22 *Osprey* VTOL tiltrotors; CH-53E *Sea Stallion* helicopters; UH-1Y *Huey* helicopters; AH-1Z *Super Cobra* helicopters; MH-60S *Seahawk* helicopters.

Date Deployed: Scheduled for delivery to the fleet in 2013

General Characteristics, *Wasp* Class

Builder: Northrop Grumman Ship Systems Ingalls Operations, Pascagoula, Miss

Power Plant: (LHDs 1-7) two boilers, two geared steam turbines, two shafts, 70,000 total shaft horsepower; (LHD 8) two gas turbines, two shafts; 70,000 total shaft horsepower, two 5,000 horsepower auxiliary propulsion motors

Length: 844 feet (253.2 meters)

Beam: 106 feet (31.8 meters)

Displacement: LHDs 1-4: 40,650 tons full load (41,302.3 metric tons) LHDs 5-7: 40,358 tons full load (41,005.6 metric tons) LHD 8: 41,772 tons full load (42,442.3 metric tons)

Speed: 20+ knots (23.5+ miles per hour)

Aircraft, depending on mission: 12 CH-46 *Sea Knight* helicopters; 4 CH-53E *Sea Stallion* helicopters; 6 AV-8B *Harrier* attack aircraft; 3 UH-1N *Huey* helicopters; 4 AH-1W *Super Cobra* helicopters. (planned capability to embark MV-22 *Osprey* VTOL tiltrotors)

Crew: Ships Company: 104 officers, 1,004 enlisted Marine Detachment: 1,894

Armament: Two RAM launchers; two NATO *Sea Sparrow* launchers; three 20mm *Phalanx* CIWS mounts (two on LHD 5-7); four .50 cal. machine guns; four 25 mm Mk 38 machine guns (LHD 5-7 have three 25 mm Mk 38 machine guns).

Date Deployed: July 29, 1989 (USS *Wasp*)

General Characteristics, *Tarawa* Class

Builders: Ingalls Shipbuilding, Pascagoula, Miss.

Power Plant: Two boilers, two geared steam turbines, two shafts, 70,000 total shaft horsepower

Length: 820 feet (249.9 meters)
Beam: 106 feet (31.8 meters)
Displacement: 39,400 tons (40,032 metric tons) full load
Speed: 24 knots (27.6 miles per hour)
Aircraft, depending on mission: 12 <u>CH-46 *Sea Knight*</u> helicopters; 4 <u>CH-53E *Sea Stallion*</u> helicopters; 6 <u>AV-8B</u> *Harrier* attack aircraft; 3 <u>UH-1N *Huey*</u> helicopters; 4 <u>AH-1W *Super Cobra*</u> helicopters

Amphibious Transport Dock—LPD

Updated: 28 February 2005

Description: Amphibious transports are warships that embark, transport, and land elements of a landing force for a variety of expeditionary warfare missions.

Features: The amphibious transports are used to transport, and land Marines, their equipment and supplies by embarked air cushion or conventional landing craft or amphibious vehicles, augmented by helicopters or vertical take off and landing aircraft in amphibious assault, special operations, or expeditionary warfare missions.

Background: The versatile *Austin*-class LPDs provide substantial amphibious lift for Marine troops and their vehicles and cargo. Additionally, they serve as the secondary aviation platform for Expeditionary Strike Groups. The oldest of the class turned 39 in early 2004. As the new *San Antonio*-class LPDs enter service, *Austin*-class LPDs will be decommissioned.

A contract for final design and construction of *San Antonio* (LPD 17), the lead ship in the class, was awarded in December 1996; actual construction commenced in August 2000. The lead ship contract included options for *New Orleans* (LPD 18), *Mesa Verde* (LPD 19) and *Green Bay* (LPD 20). The options for LPD 18 and 19 were exercised in December 1998 and February 2000. A negotiated modification added LPD 20 in May 2000. In November 2003, the Navy awarded the contract to build *New York* (LPD 21). The bow stem of *New York* was cast in 2003 using tons of steel salvaged from the World Trade Center and the keel was laid in September 2004.

In 2004, Northrop Grumman Ship Systems launched *New Orleans* and *Mesa Verde* and laid the keel for *New York*. Also the Secretary of the Navy named *San Diego* (LPD 22), *Anchorage* (LPD 23), *Arlington* (LPD 24) and *Somerset* (LPD 25). *New York*, *Arlington* and *Somerset* honor those who died in the terrorist attacks of September 11, 2001. In 2005 the Navy expects to start construction of *San Diego* (LPD 22), expects to launch *Green Bay* (LPD 20), and will commission *San Antonio* (LPD 17), first ship of the class.

The ships of the LPD 17 class are a key element of the Navy's seabase transformation. Collectively, these ships functionally replace over 41 ships (LPD 4, LSD 36, LKA 113, and LST 1179 classes of amphibious ships) providing the Navy and Marine Corps with modern, seabased platforms that are networked, survivable, and built to operate with 21st century transformational platforms, such as the MV-22 *Osprey*, the Expeditionary Fighting Vehicle (EFV), and future means by which Marines are delivered ashore.

Point of Contact:
Public Affairs Office
Naval Sea Systems Command
Washington, D.C. 20362

General Characteristics, *San Antonio* class

Builders: Northrop Grumman Ships Systems, with Raytheon Systems Corporation and Intergraph Corporation.

Power plant: four sequentially turbocharged marine Colt-Pielstick Diesels, two shafts, 41,600 shaft horsepower

Length: 684 feet (208.5 meters)

Beam: 105 feet (31.9 meters)

Displacement: Approximately 24,900 long tons (25,300 metric tons) full load

Speed: in excess of 22 knots (24.2 mph, 38.7 kph)

Aircraft: Launch or land two CH53E *Super Stallion* helicopters or two MV-22 *Osprey* tilt rotor aircraft or up to four CH-46 *Sea Knight* helicopters, AH-1 or UH-1 helicopters

Armament: Two Bushmaster II 30 mm Close in Guns, fore and aft; two Rolling Airframe Missile launchers, fore and aft.

Landing Craft/Assault Vehicles: Two LCACs or one LCU; and 14 Expeditionary Fighting Vehicles

Dock Landing Ship—LSD

Updated: 1 October 2003

Description: Dock Landing Ships support amphibious operations including landings via Landing Craft Air Cushion (LCAC), conventional landing craft and helicopters, onto hostile shores.

Background: These ships transport and launch amphibious craft and vehicles with their crews and embarked personnel in amphibious assault operations.

LSD 41 was designed specifically to operate LCAC vessels. It has the largest capacity for these landing craft (four) of any U.S. Navy amphibious platform. It will also provide docking and repair services for LCACs and for conventional landing craft.

In 1987, the Navy requested $324.2 million to fund one LSD 41 (Cargo Variant). The ship differs from the original LSD 41 by reducing its number of LCACs to two in favor of additional cargo capacity.

Point of Contact:
Public Affairs Office
Naval Sea Systems Command
Washington, DC 20362

General Characteristics, *Harpers Ferry* Class

Builders: Avondale Industries Inc., New Orleans, La.
Power Plant: Four Colt Industries, 16 Cylinder Diesels, two shafts, 33,000 shaft horsepower
Length: 609 feet (185.6 meters)
Beam: 84 feet
Displacement: 16,708 tons (16,976.13 metric tons) full load
Speed: 20+ knots (23.5+ miles per hour)
Landing Craft: Two <u>Landing Craft, Air Cushion</u>

Landing Craft, Air Cushioned

Updated: 30 December 2003

QuickTime movie of LCAC "flying" across the water during *Kernel Blitz '97*. (This movie is 1.232 MB)

Description: Air cushion craft for transporting, ship-to-shore and across the beach, personnel, weapons, equipment, and cargo of the assault elements of the Marine Air-Ground Task Force.

Features: The Landing Craft Air Cushion (LCAC) is a high-speed, over-the-beach fully amphibious landing craft, capable of carrying a 60–75 ton payload. It is used to transport the weapons systems, equipment, cargo and personnel of the assault elements of the Marine Air-Ground Task Force from ship to shore and across the beach. LCAC can carry heavy payloads, such as an M-1 tank, at high speeds. The LCAC payload capability and speed combine to significantly increase the ability of the Marine Ground Element to reach the shore. Air cushion technology allows this vehicle to reach more than 70 percent of the world's coastline, while only about 15 percent of that coastline is accessible by conventional landing craft.

Background: Concept Design of the present day LCAC began in the early 1970s with the full-scale Amphibious Assault Landing Craft (AALC) test vehicle. During the advanced development stage, two proto-types where built. JEFF A was designed and built by Aerojet General in California. JEFF B was designed and built by Bell Aerospace in New Orleans, Louisiana. These two craft confirmed the technical feasibility and operational capability that ultimately led to the production of LCAC. JEFF B was selected as the design basis for today's LCAC.

The first LCAC was delivered to the Navy in 1984 and Initial Operational Capability (IOC) was achieved in 1986. Approval for full production was granted in 1987. After an initial 15-craft production competition contract was awarded to each of two companies, Textron Marine and Land Systems (TMLS) of New Orleans, La., and Avondale Gulfport Marine, TMLS was selected to build the remaining craft. A total of ninety-one LCAC have now been built. The final craft, LCAC 91, was delivered to the U.S. Navy in 2001. This craft served as the basis for the Navy's LCAC Service Life Extension Program (SLEP). To date three operational craft have been delivered to the Navy in the SLEP configuration.

LCAC first deployed in 1987 aboard *USS Germantown* (LSD 42). LCAC are transported in and operate from all amphibious well deck ships including LHA, LHD, LSD and LPD. The craft operates with a crew of five.

In addition to beach landing, LCAC provides personnel transport, evacuation support, lane breaching, mine countermeasure operations, and Marine and Special Warfare equipment delivery.

Program Status: All of the planned 91 craft have been delivered to the Navy. A Service Life Extension Program (SLEP) is currently in progress to add service life to the craft design life of 10 years, delaying the need to replace these versatile craft.

Point of Contact:
Public Affairs Office
Naval Sea Systems Command
Washington, DC 20362

General Characteristics

Class: LCAC 1

Builder: Textron Marine and Land Systems/Avondale Gulfport Marine

Power Plant: 4-Allied-Signal TF-40B gas turbines (2 for propulsion/2 for lift); 16,000 hp sustained; 2-shrouded reversible pitch airscrews; 4-double-entry fans, centrifugal or mixed flow (lift) / 4-Vericor Power Systems ETF-40B gas turbines with Full Authority Digital Engine Control (FADEC) (2 for propulsion/2 for lift); 16,000 hp sustained; 2-shrouded reversible pitch airscrews; 4-double-entry fans, centrifugal or mixed flow (lift)

Length: 87 feet 11 inches (26.4 meters)

Beam: 47 feet (14.3 meters)

Displacement: 87.2 tons (88.60 metric tons) light; 170-182 tons (172.73 - 184.92 metric tons) full load

Range: 200 miles at 40 kts with payload/300 miles at 35 kts with payload

Speed: 40+ knots (46+ mph; 74.08 kph) with full load

Load Capacity: 60 tons/75 ton overload (54.43/68.04 tonnes)

Military lift: 24 troops or 1 MBT

Crew: Five

Armament: 2-12.7mm MGs. Gun mounts will support: M-2HB .50 cal machine gun; Mk-19 Mod3 40 mm grenade launcher; M-60 machine gun

Radars, Navgation: Marconi LN 66; I band/Sperry Marine Bridge Master E

Date Deployed: 1982

3

Command and Control

Learning Objectives

At the end of this lecture the student will be able to:

Define Command and Control and describe the three components of which form the basis of the command and control system.

Explain the purpose for "Rules of Engagement."

Describe the Composite Warfare Commander (CWC) concept, how and why it works, and the standard organization.

State the mission and functions of CIC.

Describe the four conditions of readiness.

Identify basic NTDS symbology.

Interpret the meaning of warning, weapons, and engagement orders.

Describe the Detect-to-Engage Sequence

Introduction

"The effectiveness of American Seapower depends directly on the effectiveness of the exercise of command, control, and coordination of our Naval Forces by Naval Commanders, and the means through which this exercise is accomplished."

—ADM Arleigh A. Burke, USN

Command and Control is the exercise of authority and direction by a properly designated commander. As a system, naval command and control has three components—command and control organization, information, and command and control support. *Command and control organization* encompasses the commander and the chain of command that connects superior commanders with subordinate commanders. *Information* is the lifeblood of the entire command and control system. *Command and Control support* is the structure by which the naval commander exercises command and control. It includes the people, equipment, and facilities that provide information to commanders and subordinates. In this chapter we will explore the first two elements: Organization and information. In the next chapter we will investigate the systems and methods used to support command and control.

Organizing for Command and Control

Chain of Command

The commander exercises command by issuing orders to subordinate units through the chain of command which descends directly from him to his immediate subordinate commanders, whom he holds individually responsible for the performance of their units.

Bypassing the normal channels of command is resorted to only in critically urgent situations. Orders should be originated and disseminated in time to permit subordinate commanders the maximum time available to prepare their units for the operation. Commanders must anticipate the delays involved in the successive dissemination of orders.

A critical part of any chain of command is a clear understanding of where the unit you belong to fits into that chain of command. For all fleet units there are two chains of command the operational chain of command and the administrative chain of command.

ADMINISTRATIVE CHAIN OF COMMAND
The administrative chain of command is responsible for the direction or exercise of authority over subordinate units in respect to matters such as personnel management, supply, services, maintenance, training and certification, and other matters not directly related to the conduct of operational missions. (e.g., COMNAVSURFPAC, COMCARSTRIKEGRU ONE, COMDESRON SEVEN) The administrative chain of command is below:

    A. *President of the United States*
    B. *Secretary of Defense*

C. *Secretary of Navy*

D. *Chief of Naval Operations*

E. *Component Commanders*—responsible for the administrative control of all fleet elements assigned to their geographic area (e.g. COMLANTFLT, COMCPACFLT). COMNAVEUR and COMNAVCENT are component commanders responsible for administrative control of U.S. Naval operating shore activities in Europe and the Central Command AOR. They have no administrative control over fleet elements.

F. *Type Commanders*—establish policy, control funds, and perform practically all administrative functions in their respective warfare specialties.

   (1) Ships: COMNAVSURFLANT and COMNAVSURFPAC
                ** COMNAVSURFLANT is subordinate to COMNAVSURFPAC **

   (2) Air: COMNAVAIRLANT and COMNAVAIRPAC
                ** COMNAVAIRLANT is subordinate to COMNAVAIRPAC **

   (3) Subs: COMNAVSUBLANT and COMNAVSUBPAC
                ** COMNAVSUBPAC is subordinate to COMNAVSUBLANT **

G. *Group Commanders*—responsible to TYCOM for administrative control of similar types of fleet elements (e.g., carrier, cruisers, squadrons, destroyers, etc.) in home port areas.

   EXAMPLES: Carrier Strike Groups (ex COMCARGRUs and Expeditionary Strike Groups (ex Amphibious Readiness Groups)

H. *Squadron Commanders* – responsible to Group Commanders for administrative control of a squadron of similar ship types.

   EXAMPLES: COMDESRON 12   (Commander Destroyer Squadron 12)

                COMPHIBRON 12 (Commander Amphibious Squadron 12)

                CAG 17   (Commander Carrier Air Wing 17)

I. *Unit Commanders*—are the ship's commanding officers responsible to squadron commanders for administrative control of their ship, or squadron commanding officers responsible to airwing commanders for administrative control of their squadrons.

   EXAMPLES: USS PREBLE (DDG 88)

                Strike Fighter Squadron One Three One—VFA 131 Wildcats

Bridge 1990s. WHO ARE YOU GONNA CALL? To get the latest information about his amphibious task force or enemy movement hundreds of miles away, Rear Admiral D.R. Conley can call the 7th Fleet Tactical Flag Command Center (TFCC). To carry out a plan of action, he can call the Supporting Arms Communication Center (SAC). (*Photo by PH2 Clayton Farrington, USN*)

OPERATIONAL CHAIN OF COMMAND

The operational chain of command executes those functions of command involving the composition of subordinate forces, the assignment of tasks, the designation of objectives, and the authoritative direction necessary to accomplish specific operational tasks and missions. (e.g., COMPACFLT, CTF 75, CTG 75.1) Members in the operational chain of command include:

A. *President of the United States* through the Secretary of Defense with The Joint Chiefs of Staff acting in an advisory capacity. This is referred to as the National Command Authority (NCA).

B. *Unified and Specified Commanders* (EUCOM, CENTCOM, USPACOM). See Combatant commands below for more information.

C. *Component Commanders*—responsible to unified commanders in their geographical area for the tactical employment of naval forces in their assigned geographical region. (COMLANTFLT/COMPACFLT)

D. *Numbered Fleet Commanders* (2ND, 3RD, 5TH, 6TH, 7TH)

E. *Designated Task Force Commanders* (TF60, TF75, etc.) (First number of fleet assigned) (Used for multi-Battle Group forces)

F. *Task Group Commander* (TG60.1, TG75.2, etc.) Strike Group Commanders.

G. *Task Unit Commander* (TU60.1.1, TU75.2.3, etc.) Warfare Commanders.

H. *Task Element Commander* (TE60.1.1.2, TE75.2.3.2, etc.) An individual ship or group of ships with a special purpose like a SAU or a SAG.

Rules of Engagement

Responses to certain situations are already spelled out by the Unified Commanders, responsible for a particular theater of action, in the form of Rules of Engagement (ROE) and Pre-planned Responses (PPRs). ROE are the situational criteria, usually written and promulgated for a specified area or event, that a contact must meet prior to being classified as hostile, targeted or fired upon. Often ROE includes standard warnings and procedures for identification of unknown contacts. ROE will change to meet the circumstances of different operating areas. Therefore, the criteria an unknown aircraft approaching your unit must meet to be classified as hostile in the Adriatic may be quite different from those required in the Arabian Gulf, or off the coast of Korea.

The Composite Warfare Commander (CWC) Concept

The purpose of the CWC concept is to provide *decentralized* command and control of the force. The fast-paced environment of modern warfare with its high speed aircraft, missiles, and submerged threats, makes it impossible for one person to direct the entire operation. For this reason warfare commanders are assigned to control specific areas while the CWC *commands by negation* (i.e., over-ruling an order by a warfare commander rather than overseeing every little detail). The CWC can, therefore, maintain the "big picture" without getting bogged down in any one specific action area.

The CWC will assign assets to his warfare commanders based on availability and capabilities. A multi-mission ship or airwing will typically be under control of more than one warfare commander. Each warfare commander is responsible for dispositions and employment of assets as well as generating reporting procedures and pre-planned responses to threats.

DECENTRALIZED COMMAND

Decentralization of command is accomplished by delegating control of specific warfare areas and function among principal warfare commanders, functional commanders, and coordinators. The difference between principal warfare commanders and functional commanders and supporting coordinators is important. When authorized by the CWC, warfare commanders have tactical control of resources assigned to them and may autonomously initiate action. Supporting coordinators execute policy, but do not initiate autonomous actions.

COMMAND BY NEGATION

Command by negation can be thought of as a "Command Override" option available to the CWC should he not desire a chosen autonomous action be executed. When so authorized, warfare commanders may initiate autonomous action. However, a key element in this command & control picture is nearly continuous reporting of actions and intentions of the subordinate warfare commanders. Should the CWC not agree with a chosen course of action, he will re-direct the warfare commander's efforts or tailor them to the situation through his exercising Command by Negation.

FUNCTIONING RELATIONSHIPS

OFFICER IN TACTICAL COMMAND (OTC)
The OTC has overall responsibility for successfully accomplishing the mission of the force and retains responsibility for maneuvering the main body.

COMPOSITE WARFARE COMMANDER (CWC)
The CWC is a central command authority who directs the force in a manner best suited to the tactical situation. Normally the OTC and the CWC are the same.The CWC is usually a Rear Admiral in command of a Carrier Battle Group and is usually stationed on board the CV with staff.

WARFARE COMMANDERS
Warfare Commanders may be individual ship CO's or embarked staff (ex: COMDESRON). They report to the CWC all functions concerning conduct of their respective warfare area. Each warfare commander is in charge of all weapons and sensors on all assigned assets with respect to their warfare area.

AW   AIR WARFARE COMMANDER (AWC)—The AWC is responsible to the CWC for protection of the force against hostile air threats and offensive employment of air assets (normally the cruiser CO).

AZ   SEA COMBAT COMMANDER (SCC)—The SCC is responsible to the CWC for protection of the force against hostile submarines and offensive employment of submarine assets, as well as for protection of the force against hostile surface threats and offensive employment of surface assets (Normally the DESRON Commodore). The SCC typically designates an Undersea Warfare Commander (USWC) and a Surface Warfare Commander (SUWC) to directly oversea those mission areas.

AP   STRIKE WARFARE COMMANDER (STWC)—The STWC is responsible for all power projection strikes against foreign shores (nNormally the CVW commander).

AQ   COMMAND AND CONTROL WARFARE COMMANDER (C2WC)—The C2WC is responsible for the conduct of information operations by and against the strike group as well as the maintenance of the strike group information network (normally the CV/N or LHA/D CO).

The Role of Information

Two Types of Information

Information is absolutely essential to effective Command and Control. There are two basic types of information. *Image-building information* is used to help create an understanding of the situation in order to make a decision. For instance if the CWC is to assign assets properly to subordinate commanders, he or she must have a certain amount of information in order to make good assignments. *Execution information* is used as a means of coordinating actions in the execution of a plan after a decision has been made. In this case the CWC has assigned assets and now the subordinate commanders must use new information to properly employ the assets.

Quality of Information

There are six attributes used to describe the quality of information. Although by no means all-inclusive, these characteristics provide a basis for qualitative assessment:

- *Relevance*—Information that applies to the mission, task, or situation at hand.

- *Accuracy*—Information that conveys the true situation.
- *Timeliness*—Information that is available in time to make decisions.
- *Usability*—Information that is in common, easily understood formats and displays.
- *Completeness*—All necessary information required by the decision maker.
- *Precision*—Information that has the required level of detail or granularity.

Command and Control Support

Effective command and control support helps the naval commander unify the force in the face of disorder (Fog of War) and shape the course of events to achieve a specific goal. Further, it helps the commander function effectively across the full range of conflict, in any environment and helps generate a rapid tempo of operations, while coping effectively with disruptions created by the enemy. Moreover, although our philosophy of command and control is based on our warfighting needs, it applies equally to successful mission accomplishment during operations other than war.

Combat Direction Center (CDC)

The principal shipboard facility for command and control support is called Combat Direction Center (CDC). Modern warfare requires the handling of vast amounts of information and fast reaction. To this end, CDC acts as the ship's tactical "nerve center" where most major decisions in fighting are made by the Commanding Officer or, in his stead, the Tactical Action Officer (TAO). CDC is the place where the requirements and orders of the CWC and warfare commanders are translated into action for that individual ship. The E2-C Hawkeye provides both the Strike Warfare and Air Warfare Commanders an airborne CDC as well.

Mission of CDC

The mission of a ship's Combat Direction Center is to *provide command and control stations with tactical and strategic information correlated from all sources to enable the Commanding Officer to determine the proper course of action in a multi-threat environment.*

Primary Function of CDC

The primary function of CDC is *information control and handling, which involves collecting, processing, disseminating and protecting pertinent tactical information.*

- *Collecting*
  Gathering and formatting data for processing.
- *Processing*
  Filtering, correlating, fusing, evaluating, and displaying data to produce image-building information required for commanders to take appropriate action.
- *Disseminating*
  Distributing image-building or execution information to appropriate locations for further processing or use.
- *Protecting*
  Guarding our information from an adversary's attempts to exploit, corrupt, or destroy it.

Secondary Functions of CDC

The secondary functions of CDC are to support and assist.

- *Support functions*
  (1) Radar reporting
  (2) Communications control
  (3) Aircraft control
  (4) Control of small boats and landing craft
  (5) AW/USW/SUW
  (6) NSFS
- *Assist functions*
  (1) Surface and air contact tracking and reporting. This includes computing contact course and speed, identification, and recommendations for maneuvering in accordance with the Rules of the Nautical Road for surface contacts.
  (2) Maintaining a current navigational plot for ocean transits. CDC's DR plot is updated from the Navigator's electronic and celestial fixes.
  (3) Search and Rescue: CDC coordinates all information received and provides the OOD with course recommendations, communications, and coordination with other units involved in any search and rescue effort.

(4) Tactical maneuvering: CDC provides a back-up for signal coding/decoding, recommendations to the OOD for proper maneuvering, communications log of all tactical signals.

(5) Low Visibility Piloting: CDC directs the movements of the ship based on an accurate radar navigation plot during periods of rain, fog, snow, etc. The OOD is still responsible for the safety of the ship and must balance CDC's recommendations with what he can see and hear on the bridge.

CDC Organization

CDC organization involves the assignment and utilization of personnel to accomplish specialized evolutions consistent with information handling and control/assist functions. Personnel billets are assigned by the Ship's Manning Document, and watch stations are assigned by the Watch Quarter and Station Bill, which provides details on each individuals watch assignments and responsibilities. The CDC watch organization fluctuates, along with the rest of the ship's watch organization, in accordance with the following readiness conditions:

(1) Condition I—Combat posture in the multi-threat environment (condition of maximum readiness, GQ).

(2) Condition II—Modified general quarters—used to permit some relaxation among the crew during combat readiness conditions. Condition II may also be used to set battle stations for a specific threat (eg. IIAS is set when there is only a submarine threat)

(3) *Condition III*—Wartime cruising. Extended Defensive Profile (man all major sensor and weapons systems with less than maximum readiness, i.e., one gun mount manned out of two). This allows for rapid response in a lower threat environment.

(4) *Condition IV*—Peacetime cruising. Ensure safe navigation

CDC Manning

The following are major stations which are manned in CDC during Condition IV:

(1) CDC Watch Officer—Coordinates the entire CDC team and ensures information is properly evaluated and disseminated.

(2) CDC Watch Supervisor—A senior enlisted petty officer supervising all information and watch station performance.

(3) Radar/NTDS Operators—Control radar repeaters and report contact range/bearing/altitude information.

(4) Plotters–Maintain current displays or plots on air, surface (maneuvering board), geographic (DRT), strategic and formation plots.

(5) Status Board Keepers—Maintain current displays of data collected and ensure this data is continually updated.

(6) R/T and S/P Phone Talkers—Receive and transmit information as directed.

(7) Navigation Plotters—Maintain navigation plot on same scale chart as bridge using electronic measures: radar, GPS, LORAN, DR.

Condition I, II, and III require additional manpower as weapons consoles be manned in order to defend against a possible threat. Manned weapons systems require a greater level of supervision. The primary supervisor in Combat during Condition I, II, and III is the Tactical Action Officer (TAO), the Commanding Officers direct representative in CDC tasked with "fighting the ship."

Naval Tactical Data System (NTDS)

The Navy Tactical Data System (NTDS) is a critical part of effective command and control support systems, allowing for real-time information flow between all platforms having NTDS capabilities. The system transmits target position, designation (Track Number), and weapons assignment status to all other NTDS platforms. This type of information link allows earlier detection (because you are not limited to the range or your own sensors), thus allowing more time for the CWC, Warfare Commanders, CO's and TAO's to make a sound tactical decision in the minimum time. Also, by indicating whether another platform has engaged a target, it helps prevent a duplication of effort, freeing other weapons systems for other targets.

NTDS is one of the key functional capabilities which allows the Composite Warfare Concept to work. It allows automatic transmission of information from one ship or aircraft to another without human interface. The common radar picture allows every NTDS capable platform to see what every other NTDS platform sees. NTDS is a high speed time sharing system controlled by the NTDS computer on one unit known as the NET CONTROL SHIP (NCS). The NCS interrogates each unit (known as a picket) sequentially and that unit sends its data to all others. The net cycle time, or time it takes to interrogate all Participating Units (PU) depends on the number of PUs, but in no case will exceed 4 seconds. If a unit fails to respond to an interrogation because of communications or computer failure, the next ship is then interrogated and the missed unit will not be interrogated again until the next cycle. Should the NCS unit's computer fail, a predesignated alternate will switch into the NCS mode and assume duties as NCS. There can only be one NCS at any given time.

Track Numbers

All NTDS symbols are assigned track numbers within the system to allow for a common method of tracking the status. Track numbers are octal numbers assigned by the computer (0000-8888) to each NTDS symbol.

Symbology

NTDS symbology allows every ship to classify a contact as hostile, friendly, or unknown. Further subclassification as surface, subsurface, or air aids the CWC, Warfare Commanders, CO, and TAO in determining at a glance the tactical situation.

Speed Leaders

Each moving symbol will have a line coming from the origin of the symbol indicating the relative direction speed of movement. Speed leaders of air contacts indicate the distance traversed in one (1) minute. The magnitude of surface and subsurface contacts shows the distance traveled in three (3) minutes. On fleet consoles, there is an option allowing you to adjust speed leaders to reflect movement for any period of time up to 30 minutes. This is used mostly to time air intercepts.

Data Links

1. Link 11—(TADIL A)—Two-way link between NTDS link-capable units (could be surface, air or sub). Data is displayed on NTDS consoles.
2. Link 14—One-way broadcast from an NTDS unit to non-NTDS units. Data is displayed on a teletype print out that gives the track

grid position, course, speed, classification, and engagement status. When received, this data must then be plotted on a display (VP for air, DRT/NC-2 for surface/subsurface) for it to be relevant to the TAO.

3. LINK 4A—(TADIL C)—Two-way data link between NTDS ship and interceptor aircraft. Along with data exchange, engagement and intercept orders may be transmitted to the aircraft via Link 4A.

4. LINK 16—(TADIL J)—Identical in purpose to LINK 11 and LINK 4A; however, it provides significant improvements such as node-lessness, jam resistance, flexibility of communication operations, separate transmission and data security, increased numbers of participants, increased data capacity and secure voice capability.

Coordination and Target Engagement

Warfare Commanders coordinate target engagement through issuance of *warning orders*, weapons control orders and engagement orders.

- **Warning Orders**
  Warning White – attack is improbable, all clear
  Warning Yellow – attack is probable; hostile units enroute
  Warning Red – attack is imminent; hostile units are inbound.
- **Weapons Control Orders**
  Weapons Safe – Do not shoot unless in self defense.
  Weapons Tight – Fire on contacts classified as hostile only.
  Weapons Free – Fire on any contact not identified as friendly.
- **Engagement Orders**
  Cover – Maintain fire control solution on specified contact.
  Take – Engage specified contact.
  Cease fire – Do not fire.
  Hold fire – Do not fire; destroy missiles in flight.

Warning and Weapon Control Orders are normally given for each warfare area by the warfare commander. For example "Air Warning Yellow, Weapons Tight," would translate into: an air attack is probable, fire on hostile contacts only." As an engagement ensues the warfare commander would issue an engagement order such as, "Take track 4463 with birds" to the individual unit most capable of engaging that threat. The CWC would receive reports from each warfare commander and monitor and control actions via communications networks and automated tactical data systems.

Detect-to-Engage Sequence

The "detect-to-engage" sequence provides a standard routine for engaging contacts. Regardless of the complexity of the threat or combat system, the following general steps are common to all warfare area "detect-to-engage." sequences.

(1) *Detection*—The initial recognition to the presence of a contact. Each contact will be assigned a track number to assist in identification.

(2) *Entry*—The subsequent process of entering the detected track information into the Naval Tactical Data System (NTDS).

(3) *Tracking*—The process of accurately determining and predicting target position. The sensors used to detect a track may also provide update data for the tracking operations.

(4) *Identification*—Determination on whether a track is friendly or hostile. The Identification Friend or Foe (IFF) equipment provides for positive identification of friendlies, other sensors and data are correlated to identify contacts, hostile.

(5) *Threat evaluation*—The process of determining the relative degree of threat (threat priority), that the track presents to ownship and /or vital areas/ships. Threat priority is based on the position of the track, projected closest point of approach, identification, range of weapon delivery, and time remaining to effectively engage.

(6) *Weapons pairing*—The process of assigning the optimum weapon for a given threat based on it relative threat priority, and available assets.

(7) *Engagement*—The employment of weapons to combat a threat.

(8) *Engagement Assessment (Battle damage assessment)*—The process of monitoring weapon return information and data from ship sensors to determine the success of an engagement.

Review Questions

1. What are the fours steps for processing and handling information?
2. List three support functions provided by CDC.
3. How does CDC support the Officer of the Deck?
4. Which Condition of Readiness is set at General Quarters?
5. What are the duties and responsibilities of the CDC Watch Supervisor?

6. What does the following NTDS symbol represent?
7. What does "warning order yellow" "weapons tight" mean?
8. In which phase of the DTE sequence are weapons selected?
9. State the Operational Chain of Command.
10. What is Operational Control (OPCON)?
11. Who exercises Tactical Control (TACON)?
12. Who establishes Rules of Engagement and what is there purpose?
13. What is the concept of leadership employed by the CWC?
14. What is image-building information used for?
15. State four characteristics used for qualitative evaluation of information.

Suggested Further Reading

Naval Doctrine Publication 6, *Naval Command and Control*. CNO, Washington D.C., 19 May 1996.

NAVEDTRA 10776-A: *Surface Ship Operations*, Naval Education and Training Command Government Printing Office, Washington D.C.

SWOSCOLPAC, *Engagement Systems*, Surface Warfare Officers School Division Officer Course, June 1991.

4

The Commander's Role in Developing Rules of Engagement

Lieutenant Colonel James C. Duncan, U.S. Marine Corps

SINCE PROPERLY CRAFTED RULES OF ENGAGEMENT (ROE) are essential to the success of any operation, the importance of the commander's role in this process cannot be overstated.[1] For each commander, ROE represent an integral part of command and control, and they provide the most effective means of implementing the political goals of civilian leadership, as well as the strategic decisions made by higher headquarters. Effective ROE must be flexible, and they must evolve with the operation. Because commanders are responsible for everything that their forces do or fail to do, they must take care that appropriate direction on the use of force is incorporated into ROE for the guidance of military members placed in harm's way. As in the past, future commanders can expect to face intense pressure to come up with the "right" ROE for their specific operation. To meet this challenge, commanders must be proactive in organizing their staffs so that these command responsibilities are met.

The Joint Chiefs of Staff (JCS) guidance to United States military commanders on the use of force is known as the "Standing Rules of Engagement," and it replaced the JCS "Peacetime Rules of Engagement" on

1 October 1994.[2] The current version of the "Standing Rules of Engagement" establishes the fundamental procedures and policies for U.S. military commanders during all military conflicts, contingencies and operations. The Standing Rules of Engagement are designed to assist the commander in crafting ROE for assigned missions and to lay down policy on the use of force for self-defense to ensure the safety and survival of the commander's unit and other U.S. forces in the vicinity.

In past operations, many commanders have delegated the preparation of ROE to their staffs, primarily to the Staff Judge Advocate (SJA). This has led to an ROE development process dominated by the SJA, the level of that dominance varying from command to command. Unfortunately, the SJA's orchestration of ROE development has not encouraged the formulation of ROE closely harmonized with the command's operational plans. Despite this problem, U.S. military forces utilizing the ROE prepared primarily by their SJAs have done exceptionally well; however, the crucial question for operational commanders is whether the ROE development process can be improved. The answer to this question is a resounding yes.

ROE planning for any operation should be done concurrently with the actual planning for the specific mission. Commanders must ensure that ROE are not prepared in isolation from operational planning, a dichotomy that could have disastrous consequences. Developing the "right" ROE requires active participation by a number of officers within the commander's staff. As operations progress, small and overworked staffs can have a negative impact on the quality of ROE. Commanders should prepare for the long haul by making certain that the headquarters staff is capable of a sustained battle-staff rhythm. Of equal importance is the role of the SJA in ROE development. Commanders need the active participation of the SJA, but by no means should the SJA dominate ROE development.

Since the process is similar for most operations, this article will focus on the preparation of ROE for a joint task force (JTF). While the development of ROE should not control the mission, the political or operational influences behind the mission may necessitate a limitation on the level of force to be used. These influences, normally referred to as bases, may be viewed as the specific terrain to which actual ROE must conform.

What Are the Bases for the ROE?

Crafting ROE for a JTF operation requires commanders and their staffs to understand these fundamental bases, to have a good working knowledge of the Standing Rules of Engagement, and to have a firm grasp of the joint planning process. Each basis for ROE is unique, and when integrated into the ROE development process helps shape the application of military force. There are three fundamental bases for U.S. ROE: national policy, operational requirements, and the law.[3] Without an adequate understanding of each distinct basis, a command's attempts to prepare ROE will be stymied.

National Policy. Of the three bases, national policy may be the hardest to articulate. National policy is often called the political objective. As Carl von Clausewitz argued, the use of military force is simply the means of reaching a political objective:

> War is not merely an act of policy but a true political instrument, a continuation of political intercourse, carried on with other means. What remains peculiar to war is simply the peculiar nature of its means. War in general, and the commander in any specific instance, is entitled to require that the trend and designs of policy shall not be inconsistent with these means. That, of course, is no small demand; but however much it may affect political aims in a given case, it will never do more than modify them. The political object is the goal, war is the means of reaching it, and means can never be considered in isolation from their purpose.[4]

The "political object" mentioned by Clausewitz is another term for foreign policy. In short, ROE must be consistent with foreign policy objectives. For this reason, the commander and his staff need to understand U.S. foreign policy and the ramifications of that foreign policy for the military operation at hand. Usually, the foreign policy goals or national political objectives are stated in the guidance received from higher headquarters or from the combatant commander within whose area of responsibility the operation will take place.[5]

The general goal of U.S. national security policy is "to maintain a stable international environment compatible with U.S. national security interests."[6] To support this policy, the United States has formulated a global objective of deterring armed attack against its interests. For effective deterrence, one must have the ability to fight at any level of conflict; since the United States has a broad capability to use conventional weapons (including nonlethal ones) as well as nuclear weapons, it should have a very credible deterrent. If deterrence fails, the national policy of the United States permits responses that (1) are proportional to the provocation; (2) are designed to limit the scope and intensity of the conflict; (3) will discourage escalation; and (4) will achieve political and military objectives. Thus the crafter of ROE must be familiar not only with the broad objectives of U.S. foreign policy but also with the specific political objectives to be achieved or supported by a particular mission.

When the mission is unclear or the political leadership or combatant commanders have not plainly articulated their policy on the use of force for the operation, commanders must seek clarification. If the political objective or the mission changes, operational concerns must be reexamined to determine whether modification of ROE is necessary. Furthermore, if a restriction upon the use of force required for operational purposes conflicts with the mission plan, either the plan must be modified or a change to the restriction must be sought. Early identification of inadequacies in the guidance from senior commanders or the civilian leadership, such as the absence of mission clarity because of "mission creep," or any potential limitation on the use of force otherwise dictated by operational requirements, is critical to the development of ROE. By confronting these issues from the start, the commander can reduce or prevent the waste of planning time.

As we have seen, national policy objectives interact with operational requirements, and coherent ROE can be crafted only when goals and means are consistent. It is the commander's responsibility to ensure that they are.

Operational Requirements. Within the commander's staff, the primary responsibility for operational matters lies with the operations directorate, or J-3.[7] Normally, the operational requirements the J-3 sets for a mission mirror the specific planning concepts that the staff has developed regarding unit security and the express and implied taskings. Operational concerns usually focus on the following planning elements: mobilization, employment, sustainment, redeployment of the military force, and rules of engagement.

Another operational matter that may impact upon ROE concerns the types of weapon systems to be used. For instance, if political considerations require that the use of force be in some way curtailed, there may be a need to articulate specific ROE or to issue special instructions interpreting the ROE for component commanders. For this reason, the JTF commander and the J-3 must be aware of the characteristics of all weapons to be used for the mission. This is particularly important for nonlethal systems, which may require specially drafted supplemental ROE to ensure they are used only in appropriate circumstances.

One other important operational consideration is the level of threat for the geographic location of the operation. Changes to the threat should trigger reviews and, if necessary, modifications to the ROE. Complacency or inaction regarding a change in the threat is a recipe for disaster, since the established ROE may no longer be appropriate.

The Law. The third basis for ROE is the law. Here the focus is on the tenets of American domestic law and the obligations of the United States under international law, generally the law of armed conflict. The domestic law of the United States includes the Constitution, federal statutes and regulations, court decisions, and common law.[8] Although several major bodies of law, such as the law of the sea and the law of neutrality, are part of the larger body of international law and might be applicable to the preparation of ROE, the focus here will be limited solely to the law of armed conflict.

The law of armed conflict has been defined as "that part of international law that regulates the conduct of armed hostilities."[9] It includes applicable treaty law as well as customary international law—which in 1900 the Supreme Court made a part of U.S. national law.[10] Consistent with international law as a whole, the law of armed conflict is viewed as permissive in nature: a practice not prohibited either by customary international law or by treaty is permitted.

Legal issues surrounding each operation must be examined to make sure that the ROE will comply with the domestic law of the United States and the law of armed conflict. Some principles of the law of armed conflict are harder to apply than others; two of the most difficult are *necessity* and *proportionality*. These principles play a prominent role in determining when force should be used and how much. The primary purpose of ROE is to "provide implementation guidance on the application of force for mission accomplishment and the exercise of the inherent right and obligation of self-defense."[11] All commanders and their staffs must understand these two aspects—mission accomplishment and self-defense—and how to utilize ROE as a risk-management tool.

The distinction between self-defense and mission accomplishment is easily lost if it is not understood that the meanings of necessity and proportionality are significantly different depending upon whether they are used in the contexts of self-defense or mission accomplishment. Commanders must take steps to prevent errors of this kind, since they can lead to confusion within the command concerning when to use force, thereby placing those executing a mission at greater risk.

The Use of Force in Self-Defense

The right of a sovereign nation to use force in self-defense is a fundamental principle of customary international law, closely related to national independence, national existence, and freedom from outside interference or intervention. It is acknowledged in Article 51 of the United Nations Charter:

> Nothing in the present Charter shall impair the inherent right of individual or collective self-defense if an armed attack occurs against a Member of the United Nations, until the Security Council has taken measures necessary to maintain international peace and security. Measures taken by Members in the exercise of this right of self-defense shall be immediately reported to the Security Council and shall not in any way affect the authority and responsibility of the Security Council under the present Charter to take at any time such action as it deems necessary in order to maintain or restore international peace and security.

The phrase "inherent right of individual and collective self-defense" set forth includes the right of self-defense under customary international law as it existed when the United Nations Charter was written.[12] Encompassed within this concept of self-defense is the right of anticipatory self-defense:[13] when an imminent threat to a nation's safety, security, or existence arises, that nation may protect itself through the exercise of proportionate force.[14]

Under the Standing Rules of Engagement for U.S. military forces, self-defense has been divided into three main categories. The first, *national self-defense*, consists of defending "the United States, U.S. forces, and in certain circumstances, U.S. nationals and their property, and/or U.S. commercial assets."[15] The second, *collective self-defense*, is defined as defending "designated non-U.S. forces, and/or designated foreign nationals and their property, from a hostile act or hostile intent." Unlike national self-defense, collective self-defense may not be exercised below the national level.[16] The third major category is *unit self-defense*, defined as "defending a particular U.S. force element, including individual personnel thereof, and other U.S. forces in the vicinity, against a hostile act or hostile intent."[17] Under the Standing Rules of Engagement, *individual* self-defense is a subset of unit self-defense. Since the right of self-defense does indeed extend to the individual, commanders have a duty to ensure that each person under their command has received training on the principles of self-defense as articulated in the Standing Rules of Engagement.[18]

To stress the importance of self-defense, the following statement is repeated in bold print eleven times in the Standing Rules of Engagement.

These rules do not limit a commander's inherent authority and obligation to use all necessary means available and to take all appropriate action in self-defense of the commander's unit and other U.S. forces in the vicinity.[19]

The words "all necessary means available" have a special, classified meaning under the Standing Rules of Engagement.[20] Even without this definition, however, the intent is clear. In self-defense commanders may take those actions and use those weapon systems not otherwise prohibited.

To explain the phrase "all appropriate action," the Standing Rules of Engagement state that in self-defense the commander should attempt to de-escalate the situation without employing force. If the situation cannot be controlled without force, proportional force should be used to disable or destroy the imminent threat. That is, the amount of force that lawfully may be used is no more than what suffices to remove that danger. If force is needed, it should not exceed that which is required to decisively counter the hostile act or hostile intent and ensure the continued safety of U.S. forces or other protected personnel or property.[21] When the hostile force no longer represents an imminent threat, the right to use force in self-defense ends. Since the obligation and authority for self-defense is inherent in command, all commanders have a standing duty to specify when the application of force for unit self-defense is appropriate.

The elements of self-defense are necessity and proportionality. As noted above, the meanings of these principles in the self-defense context are much different from their meanings when applied under the law of armed conflict for mission accomplishment. An understanding of these two principles is crucial to ROE, for necessity and proportionality, as amplified by the Standing Rules of Engagement, will be the basis for the judgment of the commander as to what constitutes an appropriate response when acting in self-defense.[22]

Necessity in Self-Defense. The principle of necessity is the key to determining whether a lawful reason exists for the use of force in self-defense. In this context, necessity refers to the presence of imminent danger due to the actions of adverse parties, forces, or nations. Under the Standing Rules of Engagement, the necessity for self-defense may be triggered by a *hostile act* or demonstration of *hostile intent*.[23] Of these two, hostile intent has always been the more difficult to ascertain. The concept of hostile intent may be viewed as an expression of the national right of anticipatory self-defense at the unit level. An assessment of hostile intent is not based solely on objective criteria but relies in large measure on the evaluation of intelligence information about the past, present, and future activities of a potential adversary and on the experience of the decision maker. A determination of hostile intent is, therefore, largely subjective. However, every commander should be prepared after the fact to explain why he or she felt that hostile intent was present.

Proportionality in Self-Defense. In self-defense, proportionality demands that "the force used must be reasonable in intensity, duration, and magnitude, to the perceived or demonstrated threat based on all facts known to the commander at the time."[24] Proportionality in self-defense, boiled down to the basics, involves determining how much force is necessary to overcome the imminent danger created by a hostile act or demonstration of hostile intent. Although any decision regarding how *much* force is proportionate will be subjective, the goal is to apply sufficient force to handle the threat decisively—but no more than that.[25]

When the proportionality issue arises in a self-defense context, the need to use force to respond to an imminent threat has already been triggered, and the defender is facing a situation that requires a timely use of force for self-preservation. At this point, proportionality for the defender becomes a process of deciding which weapon systems will provide the force needed to counter the imminent threat. Based on an assessment of the facts, circumstances, intelligence information regarding the imminent threat, and the available weaponry, the defender must decide on the appropriate weapons. Another factor that should be considered by the defender when making this decision is how to control the level, nature and duration of that force in order to reduce or prevent injury to civilians or damage to their property.

The Use of Force for Mission Accomplishment

When force is used to accomplish a mission, it is governed by the principles of necessity and proportionality as they apply under the law of armed conflict. In contrast to the self-defense guidance in the Standing Rules of Engagement, which remains constant, ROE for mission accomplishment must be tailored to the specific needs of the mission. Mission-accomplishment ROE are prepared by modifying, where appropriate, the Standing Rules of

Engagement with supplemental measures. Various categories of supplemental measures are set forth in that document, along with the following policy, in bold type:

> Supplemental measures do not limit a commander's inherent authority and obligation to use all necessary means available and to take all appropriate action in self-defense of the commander's unit and other U.S. forces in the vicinity.[26]

As indicated, the right and obligation of self-defense always exists, whatever the supplemental measures. Supplemental measures "define the limits or grant authority for the use of force for mission accomplishment, not for self-defense."[27] Through supplemental measures the commander may either grant to subordinate units additional latitude of action or may impose specific constraints on how to carry out a mission.

Confronted with ROE that include constraints, or withhold authorities, that threaten to compromise a command's accomplishment of its assigned mission, the commander must request a change to the ROE by what is called a supplemental measure. Any supplemental measures selected or drafted must be consistent with the three bases for ROE: national policy, operational requirements, and the law.[28]

Necessity in Mission Accomplishment. In armed conflict, only that amount of force necessary to defeat the enemy may be employed; any application of force unnecessary to that purpose is prohibited. In short, necessity limits the amount and kind of force permitted to that which is authorized by the law of armed conflict.[29] For example, the unjustified killing of prisoners of war would be illegal under the law of armed conflict; therefore, the concept of necessity would prohibit the use of force for such a purpose. Necessity, or "military necessity," as it is sometimes called, connotes a limitation on the application of military force.[30] It is important to note that military necessity does not mean military expediency. Military expediency may not be used as an excuse to expand the use of force as a matter of necessity in order to sanction violations of protections set forth in the law of armed conflict. Military necessity simply permits commanders to use force to attack lawful military objectives when there is a need to do so. "Lawful military objectives," in turn, are defined as those objectives whose "nature, purpose, or use make an effective contribution to military action and whose total or partial destruction, capture, or neutralization at the time offers a definite military advantage."[31] The term "definite military advantage" is often considered the advantage gained by the neutralization of an enemy's war-fighting and war-sustaining capability.[32] Under this principle, force may lawfully be used, as necessary, against places or things that are being used for a military purpose by an adversary, or against the military personnel of that adversary.

Under the Standing Rules of Engagement, once an adversary's military units have been declared hostile by appropriate authority, U.S. military units may engage those forces (including their military equipment and sustainment structure) worldwide, except in neutral territory, *without* first observing a hostile act or a demonstration of hostile intent by that force.[33] An adversary's forces are most often declared hostile for purposes of mission accomplishment.

Proportionality in Mission Accomplishment. What constitutes proportional force under the law of armed conflict is very different from what may lawfully be used in self-defense to respond to a hostile act or to a demonstration of hostile intent. The primary difference involves the end state. In war, the goal is to obtain the submission of the adversary through the defeat of the adversary's military structure or units by overwhelming force. In contrast, self-defense merely allows the use of force to counter the threat posed by an adversary, to ensure the continued safety of one's own forces, and, where applicable, to deter or modify the future behavior of an adversary (a state or terrorist organization).

The principle of proportionality in the law of armed conflict context also requires that a military response or attack not cause damage to civilian property (collateral damage) or death and injury to civilians (incidental injury) that is *excessive* in light of the anticipated military advantage.[34] The best decision-making tool available to help the commander determine what constitutes proportionate force for mission accomplishment under the law of armed conflict is the "balancing test."[35] This test weighs the possible harmful effects of the contemplated level of force in terms of incidental injury to civilians and collateral damage to civilian property against the expected military advantage. Incidental injury of civilians or collateral damage to civilian property during an attack on a legitimate military target is lawful if the commander-having taken all reasonable precautions to minimize civilian injury and property damage consistent with the accomplishment of the mission and security of the force-can be judged to have reasonably balanced these unavoidable costs to an enemy's civilian population and property against the military advantage to be gained.

Responsibility for Crafting ROE in a JTF

For each JTF operation, the responsibility for ROE resides with the commander. However, time limitations and the multitude of tasks requiring the commander's personal attention often reduce commanders' ability to involve themselves directly in ROE development. Accordingly, JTF commanders must organize their staffs in a manner that will maximize the productivity of their staff officers and at the same time generate timely but thorough ROE.

Commanders may enhance their own efforts and those of their staffs regarding the preparation of the ROE through two organizational techniques: establishing a synergistic atmosphere within the JTF staff, and linking the ROE development process with the activities of mission planners. How well the commander integrates the ROE development process into operational planning and execution will impact directly upon the effectiveness of the operation. In sum, the challenge faced by the commander is to organize a functional procedure that allows for the development of the ROE in tandem with mission planning. The exact method will vary from commander to commander, based upon such factors as the perceived strengths or weaknesses of individual staff officers and battle-staff rhythm.

Under joint doctrine, the staff section with primary responsibility for crafting ROE is the operations directorate; fittingly, this staff section (the J-3) is assigned the principal duties of planning and conducting operations. This does not mean, however, that the J-3 should draft the ROE in a vacuum. The J-3 should be the leader of the staff effort to carry out the commander's vision of the operation, including the essential ROE development process.

Two distinct methods that a commander might use to integrate the ROE development process into mission planning are to establish an ROE cell or a joint planning group (JPG).[36] By forming an ROE cell, the commander establishes a special staff working environment that permits necessary individual and group interchanges. The ROE cell promotes these by pulling together the right staff members with the right information at the right times. Since the staff element responsible for the preparation of the ROE is the J-3, the commander should place the operations officer or deputy in charge of the ROE cell. Other members of the ROE cell should include representatives from the intelligence (J-2) and future-plans (J-5) directorates, the staff judge advocate, subject-matter experts, and officers with pertinent warfare subspecialties (such as submarine warfare or air operations). Often an expert, such as an engineer, can provide the ROE cell with a wealth of information: for example, the structural weaknesses of a target, the best weapon or explosives to destroy the target, and the possible environmental impact of a target's destruction. Since the ROE cell works best in a stable, nonfluctuating environment, it should be used primarily during the deliberate planning cycle for a contingency. In addition, the ROE cell function should be established early in the planning process, so that individual staff responsibilities and procedures may be set.

The obvious advantage gained from the interaction of the various staff members is the ability to anticipate and brainstorm. The overall quality of the ROE will be improved if the ROE cell is able to examine the effect that proposed ROE may have in existing situations and foreseeable circumstances. By using the ROE cell commanders move one step closer to their ultimate goal, which is to create clear, unambiguous guidance so that there will be no hesitation by the members of their commands when a decision must be made on when and how to use force.

Once established, the ROE cell will be able to handle a number of functions that are vital to the development of ROE. Within the JTF, the ROE cell should become the commander's focal point for interpreting ROE policy guidance from national authorities or the combatant commander, handling potential changes to the ROE based upon threat changes, drafting or reviewing supplemental requests to modify the ROE, and establishing procedures for preparing and developing ROE training packages.

One weakness of the ROE cell is that it is less effective for time-sensitive crisis action planning for the "branches and sequels" that may arise during operational planning. Crisis planning disrupts the normal functions of an ROE cell and stretches the officers in it too thin. In short, conducting crisis action planning inside the ROE cell would dramatically increase its battle-staff rhythm. For crisis action planning, the JTF commander may prefer to activate a JPG to adjust the ROE. In order to have the flexibility to handle ROE development for both deliberate and crisis action planning, the commander may decide to create an ROE cell and a JPG, but to create both a commander must have a robust and experienced JTF staff.

The JPG would become the core element for crisis action planning within a JTF. An effective JPG should be smaller in size than an ROE cell, but it should include the SJA (or his deputy) along with key members from the J-2, the J-3, and the J-5. With the staff judge advocate in the JPG, the requisite synergy will be present for the concomitant development of the ROE *and* the operational courses of action. The primary benefit of the JPG is that it permits early elimination of courses of action that cannot be supported by the underlying bases upon which the ROE must be developed. After the crisis action planning is complete and the execution phase begins, the commander should stand down the JPG and use the ROE cell for all ROE issues until the next crisis situation arises.

Throughout the operation, the commander should emphasize to the staff that ROE development is continuous and that it does not end with the execution phase of the operation. Should the mission or the threat change, the ROE must be reviewed.

For a JTF operation, commanders must organize their staffs to provide the setting most conducive for the development of ROE. One recommended approach is for the commander to create an ROE cell and, if needed, a JPG. No single staff section within the JTF should be allowed to dominate the preparation of the ROE. Commanders and the members of the ROE cell and JPG must have a firm understanding of national policy, operational requirements, and law as they apply to the JTF operation.

The commander should encourage the staff to be vigilant in identifying issues affecting the development of the ROE. Once essential issues are known, the ROE cell should review the potential supplemental measures in the Standing Rules of Engagement, or draft new ones to satisfy those issues. Development of the ROE for the operation should parallel the operational planning and the preparation of the courses of action for each mission. Any special capability the command needs should be scrutinized in light of the bases of ROE. If a conflict arises between the mission and the ROE, either the mission must be changed or the ROE must be modified.

Commanders also would be wise to ensure that the communication links within their staffs allow all information relevant to the use of force to find its way smoothly and quickly into the ROE cell (or the JPG). For example, the ROE cell (or JPG) must be provided with the special characteristics of weapon systems that may be used. This type of information would be critical, since it might influence a decision regarding what, if any, supplemental measures may be required for the proper employment of that weapon. By creating an ROE cell (and, where required, a JPG), commanders can set the stage for the generation of ROE tailor-made for their operations.

Notes

1. The term "ROE" is defined for U.S. military forces as "rules which delineate the circumstances and limitations under which United States forces will initiate and/or continue combat engagement with other forces encountered." Joint Chiefs of Staff, *Department of Defense Dictionary of Military and Associated Terms*, Joint Publication 1-02 (Washington, D.C.: 23 March 1994), p. 329.

2. Chairman, Joint Chiefs of Staff Instruction [CJCSI] 3121.01, Subj: "Standing Rules of Engagement for US Forces" 1 October 1994. CJCSI 3121.01 was modified in 1999 by CJCSI 3121.01A (1999) [hereafter "Standing Rules of Engagement"], p. 1. The "Peacetime Rules of Engagement for US Forces" were promulgated by a memorandum of the Secretary of the Joint Staff on 28 October 1988.

3. For a discussion of the ROE bases, see Richard J. Grunawalt, "The JCS Standing Rules of Engagement: A Judge Advocate's Primer," *Armed Forces Law Review*, vol. 42, 1997, p. 247, and J. Ashley Roach [Capt., JAGC, USN], "The Rules of Engagement," *Naval War College Review*, January–February 1983, pp. 47–48.

4. Carl von Clausewitz, *On War*, ed. Michael Howard and Peter Paret (Princeton, N.J.: Princeton Univ. Press, 1984), p. 87.

5. Each combatant commander receives guidance on foreign policy objectives for his area of responsibility from the National Command Authorities (NCA), the president and the Secretary of Defense. For the United States, the president has the overall responsibility to establish and implement foreign policy, as laid down in the Constitution, Art. II.

6. "Standing Rules of Engagement," p. A-3.

7. See Joint Chiefs of Staff, *Unified Action Armed Forces (UNAAF)*, Joint Publication 0-2 (24 February 1995), pp. IV-12 through IV-14. The J-3, one of the standard staff directorates, assists the JTF commander in the decision-making and execution processes of a mission. Other standard JTF directorates include the J-1 (manpower and personnel), J-2 (intelligence), J-4 (logistics), J-5 (plans and policy), and the J-6 (command, control, communications and computer [C4] systems). In addition, a JTF staff has special staff groups that "furnish technical, administrative and tactical advice and recommendations to the commander and other staff officers" (ibid., p. IV-13). Examples of special staff groups are the SJA, the medical officer, the dental officer, the comptroller, and the public affairs officer. The sole function of the members of the staff and the special staff groups is to support the JTF commander. Staff members have only the authority delegated to them by the commander. Ibid., p. IV-12.

8. Common law is defined as "those principles and rules of action . . . which derive their authority solely from usages and customs" (*Black's Law Dictionary: Definitions of the Terms and Phrases of American and English Jurisprudence, Ancient and Modern*, 6th ed. [St. Paul, Minn.: West, 1990)], p. 276). As used here, common law refers to the legal theory and customs of the United States, as distinguished from the Constitution and from statutory law that has been enacted by Congress.

9. Joint Publication 1-02, p. 215.

10. Customary international law is defined as a custom, practice or usage that has attained "a degree of regularity and is accompanied by a general conviction among nations that behavior in conformity with that practice is obligatory" (U.S. Navy Dept., *Annotated Supplement to the Commander's Handbook on the Law of Naval Operations*, NWP 1-14M/MCWP 5.2.1/COMDTPUB P5800.1 (hereafter NWP 1-14M) [Washington, D.C.: 15 November 1997], pp. 5–8). The U.S. Supreme Court ruled on customary international law in *The Paquete Habana*, 175 U.S. 677, 20 S.Ct. 290, 299 (1900).

11. "Standing Rules of Engagement," p. A-1.

12. J. L. Brierly, *The Law of Nations: An Introduction to the International Law of Peace*, 6th ed. (New York: Oxford Univ. Press, 1963), pp. 416–21; Gerhard von Glahn, *Law among Nations: An Introduction to Public International Law*, 7th ed. (Boston: Allyn and Bacon, 1996), pp. 562–5; and NWP 1-14M, p. 4–10. For a discussion of self-defense as an inherent right see Yoram Dinstein, *War, Aggression and Self-Defence*, 2d ed. (Cambridge: Grotius; New York: Press Syndicate of the University of Cambridge, 1994), pp. 179–82.

13. An excellent discussion of anticipatory collective self-defense may be found in George K. Walker, "Anticipatory Collective Self-Defense in the Charter Era: What the Treaties Have Said," in *The Law of Military Operations: Liber Amicorum Professor Jack Grunawalt*, ed. Michael N. Schmitt, Naval War College International Law Studies, vol. 72 (Newport, R.I.: Naval War College Press, 1998), pp. 365–425.

14. For a discussion of the right of anticipatory self-defense as it relates to targeting see S. V. Mallison and W. T. Mallison, "Naval Targeting: Lawful Objects of Attack," in *The Law of Naval Operations*, ed. Horace B. Robertson, Jr., Naval War College International Law Studies, vol. 64 (Newport, R.I.: Naval War College Press, 1991), p. 241. The negotiating history of Article 51 shows that necessary and reasonable anticipatory self-defense was intended to be an essential element of individual and collective self-defense. Ibid., pp. 263–4, and Stanimar A. Alexandrov, *Self-Defense against the Use of Force in International Law* (The Hague: Kluwer Law International, 1996), pp. 97–9 and 143–4.

15. "Standing Rules of Engagement," p. A-4.

16. Ibid.

17. Ibid.

18. Ibid., pp. A-4, A-6, GL-16. Under the Standing Rules of Engagement individuals have "the inherent right to use all means available and to take all appropriate action to defend themselves and other U.S. forces in their vicinity" (p. A-4).

19. Ibid., p. A-2.

20. For the classified definition of "all necessary means available," see the "Standing Rules of Engagement," p. GL-6.

21. Ibid., p. A-6.

22. Ibid., pp. A-4 and GL-24.

23. *Hostile act* is defined as "an attack or other use of force by any civilian, paramilitary, or military force or terrorist(s) with or without national designation against the United States, U.S. forces, and in certain circumstances, U.S. nationals, their property, U.S. commercial assets, and other designated non-U.S. forces, foreign nationals and their property" (ibid., pp. GL-13 and GL-14). *Hostile intent* is defined as "the threat of [the] imminent use of force against the United States, U.S. forces, and in certain circumstances, U.S. nationals, their property, U.S. commercial assets, and/or other designated non-U.S. forces, foreign nationals and their property" (ibid., p. GL-14).

24. Ibid., p. A-5.

25. For subjectivity, see Leslie C. Green, *The Contemporary Law of Armed Conflict* (Manchester, U.K.: Manchester Univ. Press, 1993), p. 331.

26. "Standing Rules of Engagement," p. J-1.

27. Ibid.

28. See Grunawalt.

29. Burleigh C. Rodick, *The Doctrine of Necessity in International Law* (New York: Columbia Univ. Press, 1928), pp. 58–9, *119.*

30. NWP 1-14M, p. 5-4. See also Green, pp. 118–9, and Geoffrey S. Corn [Maj., USA], "International & Operational Law Note (Principle 1: Military Necessity)," *Army Lawyer*, July 1998.

31. Protocol I, Art. 52 (2), reprinted in Adam Roberts and Richard Guelff, eds., *Documents on the Laws of War*, 2d ed. (Oxford, U.K.: Clarendon Press, 1989), p. 417. The United States considers this statement part of the customary international law. U.S. Defense Dept., General Counsel, letter of 22 September 1972, reprinted in *American Journal of International Law*, vol. 67, 1973, pp. 123-4.

32. See NWP 1-14M, p. 8-2.

33. In neutral territory, which includes neutral airspace, neutral water, and neutral land, all acts of hostility are prohibited (1907 Hague Convention V Respecting the Rights and Duties of Neutral Powers and Persons in Case of War on Land, 18 October 1907, chap. I, "The Rights and Duties of Neutral Powers" [hereinafter 1907 Hague Convention V], Art. 1, reprinted in Roberts and Guelff, eds., p. 63). When a neutral state is unwilling, unable, or otherwise fails to enforce its obligation to prevent unlawful belligerent use of its territory, under the law of neutrality an exception arises that allows for the engagement of belligerent forces operating within the neutral's territory by the other belligerent (Green, pp. 260-1 and NWP 1-14M, p. 7-6). For U.S. military forces, this exception is known as "self-help." For a discussion of self-help see von Glahn, pp. 529–52; Dinstein, p. 175; and Alexandrov, p. 11-9. For hostile act or intent, "Standing Rules of Engagement," p. A-5, which specifies authority in the U.S. military to declare a force hostile.

34. For collateral damage, the 1977 Geneva Protocol I Additional to the Geneva Conventions of 12 August 1949 [hereinafter Protocol I], and Relating to the Protection of Victims of International Armed Conflict, 12 December 1977, Part IV, Section I, Art. 57 (4), reprinted in Roberts and Guelff, eds., 2d ed., p. 420. See also NWP 1-14M, pp. 8-4, 8-5. For incidental injury, Protocol I, arts. 48, 49, 50, and 57 (4), reprinted in Roberts and Guelff, pp. 414–20; and Green, p. 120. See also NWP 1-14M, pp. 8-4, 8-5. For military advantage, Protocol I, Art. 57, reprinted in Roberts and Guelff, pp. 419–20.

35. For a discussion of the balancing test see Michael Bothe et al., *New Rules for Victims of Armed Conflict: Commentary on the Two 1977 Protocols Additional to the Geneva Convention of 1949* (The Hague: Martinus Nijhoff, 1982), pp. 309–11.

36. On ROE cells, Commander Dave Wagner, U.S. Navy, in a presentation on 8 August 1997 at the Naval Justice School, Newport, R.I., stated that Brigadier General M. R. Berndt (U.S. Marine Corps, Director, Joint Training Analysis and Simulation Center and the J-7, U.S. Atlantic Command) had approved the inclusion of the ROE cell concept in drafts of the Joint Tactics, Techniques and Procedures Publication. For joint planning groups, conversation with Cdr. Wagner, Joint Training Analysis and Simulation Center/U.S. Atlantic Command, 12 February 1998. See also "Standing Rules of Engagement," Enclosure L, pp. L-1 through L-4.

5

Surface Warfare

Learning Objectives

At the end of this lecture the student will be able to:

Describe the principle objectives of SUW.

State the command relationship between the CWC and the SUWC.

Describe organization of GCCS-M.

Describe the concept of a Surface Strike Group (SAG).

Describe the various platforms and weapons involved in SUW.

Describe the four phases of SUW.

Describe the considerations when using Over-the-Horizon Targeting.

Describe the problems associated with Over-the-Horizon Targeting.

Introduction

The days of close engagements between battleships are long gone and it has been proven that air supremacy is not enough to render surface ships ineffective. The improved air defense capabilities of ships coupled with cruise missile technology requires a coordinated effort between TACAIR, surface ships, and submarines to conduct effective SUW engagements.

Surface Warfare (SUW), formerly called Anti-Surface Warfare (ASUW), is the destruction or neutralization of enemy surface combatants

FIRST UNDER WAY—PORT ROYAL (CG 73), the 17th TICONDEROGA (CG47) Class Aegis guided missile cruiser, and 19th ship of the class to be built for the U.S. Navy by Ingalls Shipbuilding division of Litton in Pascagoula, Mississippi, completed initial predelivery sea trials in December 1993.

Ingalls Shipbuilding Photo

and merchant ships. *The principle objective of SUW is to deny the enemy the effective use of surface warships and cargo carrying capability.* Although there are many similar platforms and weapons in both SUW and Strike Warfare, *Surface Warfare is a sea control mission*, while Strike is a power projection mission against targets ashore.

Command Relationships

The Surface Warfare Commander, SUWC, is one of the warfare commanders subordinate to the Composite Warfare Commander. To conduct SUW the SUWC coordinates use of assets with the other warfare commanders. In a standard Carrier Battle Group setup, the SUWC is normally the CVW or Destroyer Squadron (DESRON) Commander.

The SUWC may detach a *Surface Action Group* (SAG), comprised of two or more ships, to destroy or neutralize enemy surface warships and their cargo carrying capability. The SUWC will also designate a *SAG commander*. Detachment of a SAG creates both an offensive and defensive baseline, and forces the enemy's attention, at least in part, away from the high value units. Thus, friendly forces inside the baseline may concentrate on other targets or warfare areas.

Global Command and Control System—Maritime (GCCS-M)

The Global Command and Control System—Maritime (GCCS-M) previously known as JMCIS (Joint Maritime Control Information System), is the Navy's primary fielded Command and Control System.

The objective of the GCCS-M is to satisfy Fleet requirements through the rapid and efficient development and of display capability. GCCS-M enhances the operational commander's war fighting capability and aids in the decision-making process by receiving, retrieving, and displaying information relative to the current tactical situation. GCCS-M receives, processes, displays, and manages data on the readiness of neutral, friendly, and hostile forces in order to execute the full range of Navy missions (e.g., strategic deterrence, sea control, power projection, etc.) in near-real-time via external communication channels, local area networks (LANs) and direct interfaces with other systems.

The GCCS-M system is comprised of four main variants, Ashore, Afloat, Tactical/Mobile and Multi-Level Security (MLS) that together provide command and control information to war fighters in all naval environments. GCCS-M provides centrally-managed C4I (Command, Control, Computers, Communications and Information) services to the Fleet allowing both United States and allied maritime forces the ability to operate in network-centric warfare operations.

GCCS-M is organized to support three different force environments: Afloat, Ashore and Tactical/Mobile. Afloat configurations can be categorized as force-level and unit-level configurations. Ashore configurations of GCCS-M are located in fixed site Tactical command centers designed to provide the Joint Task Force Commander with similar C4I capabilities when forward-deployed ashore. In order to allow for maximum interoperability among GCCS systems at all sites and activities, GCCS-M utilizes common communications media to the maximum extent possible. The Secure Internet Protocol Router Network (SIPRNET), Non-Secure Internet Protocol Router Network (NIPRNET) and the Joint Worldwide Intelligence Communication System (JWICS) provide the necessary Wide Area Network (WAN) connectivity. Operating "system-high" at the Secret and SCI security levels, both networks use the same protocols as the Internet. In addition to the SIPRNET operating at Secret/SCI security levels, GCCS-M supports collaborative planning at the National Command Authority level by providing Top Secret connectivity to a limited number of sites.

GCCS-M had been implemented traditionally on high-performance UNIX workstations, until recently only these platforms were powerful enough to run GCCS-M software. However, with the exponential increase in processing capability, migrating GCCS-M to the PC environment is a very practical and logical decision. Now designed for the PC environment, GCCS-M becomes largely hardware independent, meaning that it uses almost all existing hardware platforms.

SUW Platforms and Weapons

Attack Weapons

Ships and Submarines

Aircraft
   (1) TACAIR—Carrier based F/A-18 can be deck launched or airborne
       in a SUCAP (Surface Combat Air Patrol) role.
   (2) P-3C, B-52, and S-3B

Targeting Platforms

Ships and Submarines

Aircraft
   (1) S-3, P3C, E-2C

   (2) HS, HSL

Weapons

Harpoon: The Harpoon is a versatile fire-and-forget anti-ship missile. The Harpoon was initially conceived for aircraft use against surfaced Soviet Echo-class cruise missile submarines. Subsequently the missile was developed for air, surface, and submarine launch against surface targets. This missile is carried on most U.S. surface combatant classes, being launched from surface-to-air missile launchers (Mk 13), vertical launchers (Mk 41), and stand-alone canisters (Mk 141). Submarines can launch the Harpoon encapsulated from standard 21-inch torpedo tubes. The P-3C, B-52, and S-3B can carry Harpoon. Harpoon missiles have a range of 75+ nm.

Surface to Air Missiles (SAM): Used in the surface mode.
Guns: 5" 62 Mk. 45 and the 76 mm Mk. 75 Oto Melera automatic guns.
Torpedoes: MK 48 torpedoes from submarines.
Mines: Relatively cheap and very capable.
Bombs: Laser and unguided Bombs.

SUW Phase

SUW is conducted in four distinct phases. Throughout each phase, the commander must remain prepared for enemy counter-attack.
   (1) *Surface, Surveillance, Communications and Identification Phase (SSC&I)*
   (2) *Approach Phase*
   (3) *Attack Phase*
   (4) *Post Attack Phase*

SSC&I Phase

The primary objective of the SSC&I phase is to locate, identify, and target potentially hostile contacts. Location, identification, and targeting can be accomplished using either passive or active methods employed by ownship or Over-the-Horizon Targeting (OTH-T) platforms.

Passive Methods

Passive methods employ receiving sensors only, thus no emissions are vulnerable to detection. For this reason visual identification is considered a passive method. Passive methods typically come from the firing unit's own Electronic Support (ES) information, third party ES information, a combination (using ES information from multiple units), or external intelligence sources (i.e. satellites).

ES may provide position information, either through a cross fix or Target Motion Analysis (TMA), and the firing unit may be able to identify and obtain a firing solution on the enemy. Therefore, it is vital that friendly forces monitor known enemy radar and fire control frequencies, as even a momentary detection on ES systems may be enough to finalize the firing solution. Of note however, passive criteria ONLY is no longer an accepted means of targeting and current doctrine no longer supports weapons release based on ES information only.

Intelligence information may also come from a satellite or other highly classified sources. However, information from these sources often is provided too late to be of use for immediate targeting. These sources are typically more useful in providing search areas and threat warnings.

Active Methods

Active methods employ the use of radiated energy thus *units are vulnerable to detection*. The returns from this energy are analyzed to determine bearing and range. Thus active methods provide no identification.

Approach Phase

The approach phase is a four-step process.

STEP 1: *Organize SSG.*
STEP 2: *Detach SSG to prosecute.*
STEP 3: *Review target information and pass it to Fire Control Station (FCS).*
STEP 4: *Maintain emissions control (EMCON) as directed.*

Attack Phase

A successful engagement on a target with a capable anti-ship missile defense requires many missiles from different directions to arrive simultaneously. Considerations of the attack are outlined below.

(1) Determination and dissemination of time on top (TOT), simultaneous impact, for desired missiles.

(2) Missile inventory.

(3) Number of missiles required to neutralize the intended target. (Calculated estimates are available in warfare publications).

(4) The missile selects the first target it sees. Therefore, it is difficult to ensure the high value unit of an enemy formation is targeted rather than the escorts.

(5) Area of Uncertainty (AOU): If targeting information is accurate and timely, the AOU, or section of ocean where the contact could possibly be found, will be small. Conversely, older information will produce very large AOU's covering hundreds of square miles of ocean.

(6) Engagement Planning Figure of Merit (EPFOM): When an engagement plan for a contact is created, the weapons systems will calculate the probability of the weapon acquiring the intended target. This probability is based on the size of the AOU; serious consideration must be given to an engagement plan with a low EPFOM.

Post-Attack Phase

Battle Damage Assessment (BDA) must be conducted after the attack phase. This may be done using ES, radar, visual, or sonar. The BDA will determine a course of action, such as:

(1) Attack again

(2) Withdraw from the area

(3) Detach additional SSG

Over-the-Horizon Targeting (OTH-T)

OTH-T embodies the general concepts of long range targeting. Typically a forward stationed ship or aircraft transmits information back to the main body for use in fire control systems. Often referred to as pickets these forward stationed platforms employ the following methods to accomplish their objective:

(1) Single ship—ESM and TMA
(2) Multiple ship—ESM cross fix
(3) Ship and Aircraft (A/C)—ESM cross fix, A/C radar/visual

Things to Consider When Conducting OTH-T

(1) Do the units have enough angular separation to provide an accurate ES fix?
(2) Is the fix achieved through the A/C's radar or visual identification?
(3) How vulnerable will the A/C be when it elevates and radiates?
(4) Can A/C pop-up, take a few sweeps, and then drop down?
(5) Do the Rules of Engagement (ROE) require a visual identification?

Problems With OTH-T

(1) *Targeting Inaccuracies.* These may occur from the sources listed below.
   (a) Platform position inaccuracies, navigation errors.
   (b) Relative platform position inaccuracies, gridlock errors.
   (c) Sensor bearing and range inaccuracies, equipment limitations.
   (d) A/C communications/link cannot be conducted if the A/C is on the deck at a long range. The A/C will have to depart the area and then elevate. This creates a navigation error problem that includes the A/C's own sensor inaccuracies.
(2) *Missile Flight.* The following factors introduce several possibilities for error.
   (a) Time of Flight—A Mach .9 missile requires 7 minutes to fly 63 NM.
   (b) True Wind—Both cross-range and down-range components will tend to alter the flight path of the missile.
   (c) Target Course and Speed—30-knot speed combined with 180 degree alteration in course will result in a large difference from predicted position.
   (d) Missile speed—Air temperature will determine whether missiles will fly faster or slower than designed speed dependent upon air temperature. Missile range is determined by the time

of flight. If not compensated for, a missile that is flying faster than normal it will overfly the target.

(3) *Identification.* The missile has no IFF. It will pick the first contact it sees after its seeker activates.

(a) You must either set way-points for the missile or off-set the aim point of the missile so that only the intended target will be inside the seeker pattern. Consequently, a firing ship must know the location of all ships in the area.

(b) The OTC may have ROE that require a specific EPFOM for the target and for non-targets. (For example a target must have a probability of acquisition of greater than 80% and a non-target must have a EPFOM of less than 20%.) Neutral or unknown contacts in the area may inhibit shooting.

Review Questions

1. What is a SSG and how is it used?
2. What does AOU mean and what is its impact on targeting?
3. What sensors can be used to accomplish BDA?
4. What are the four phases of SUW.
5. What particular problems must the SUWC consider during OTH-T?
6. Describe some of the factors influencing employment of aircraft in SUW operations.
7. What is EPFOM?

6

Undersea Warfare

Learning Objectives

At the end of this lecture the student will be able to:
Define Undersea Warfare.
Identify the major players in Undersea Warfare.
Identify submarine roles and missions.
Describe the various platforms and weapons involved in USW.
Discuss the strengths and weaknesses of the various platforms.
Explain submarine attack philosophy.

Introduction

Undersea warfare (USW) is a functional term that better describes what was commonly known as anti-submarine warfare (ASW). The term, USW, accurately emphasizes the military operations and programs being conducted under the surface of the oceans, not just ASW. The Submarine Force is the primary USW organization, but there are several other communities that contribute to USW. The USW community includes, but not limited to, the Integrated Undersea Surveillance System (IUSS) community, Special Operation Forces (SOF), Mine Warfare community, Naval Meteorology and Oceanography Command (METOC), the ASW experts

USS *Helena* (SSN 725)

in the surface line and aviation communities, and a wide range of academic institutions, laboratories, systems centers, and other organizations. VADM Giambastiani, former COMSUBLANT, once said, "Undersea warfare is a team sport, a combined arms effort. No one community can perform the Navy's undersea warfare role by itself. The combat power of submarines is magnified by the contributions of our teammates."

The undersea warfare community is working together to improve our ability to operate effectively in a challenging medium. The focus is more than just anti-submarine warfare (ASW). Here are a few examples of other USW operations and programs in progress: unmanned underwater vehicles (UUV), Advanced Rapid Commercial Of the Shelf (COTS)

Insertion Sonar system (ARCI), Improved Submarine Launched Mobile Mine (ISLMM), and the Advanced Sea Air Land (SEAL) Delivery System (ASDS).

Asymmetric Warfare Threats

The Navy has shifted focus from an independent blue-water, open-ocean force to one capable of handling regional challenges in the littoral regions. Numerous independent, analytical studies conclude that 21st century naval warfare will be marked by the use of asymmetrical means to counter a U.S. Navy whose doctrine and force structure enable robust power projection ashore from the littorals. Asymmetric warfare implies that potential adversaries will use easily acquired weapons systems that exploit perceived weaknesses in our doctrine or capabilities.

A disturbing trend is the increasing proliferation and commercial access to technology developed and deployed by major powers. Thanks to a robust international arms market, a regional power could acquire large numbers of relatively low-cost cruise missiles, simple tactical ballistic missiles, diesel submarines, mines, and information warfare technology for a modest investment. Regional powers also have increasing access to commercial satellites capable of providing the necessary communication, command and control network as well as a detection capability that enables targeting ships at sea. Access to asymmetric systems allows regional powers and future peer competitors to build a creditable anti-access denial area and prevent or delay our Navy's use of the littorals. All have the potential to delay or reduce our Navy's ability to project power from the sea.

Diesel Submarines. Our Navy operates all over the world. Forward presence is one of the Navy's key missions. We cannot count on basing our ships in foreign countries close to deployment areas. Consequently, America has chosen to build only nuclear powered submarines because of their speed, endurance, and ordnance load. But most countries do not require forward presence and do not have the necessary budget and technology to maintain a modern nuclear submarine force. Outfitted with modern sensors, processing capability, and weapons, these diesel submarines become formidable platforms.

The following example demonstrates how difficult it is to detect and destroy a diesel submarine. During the 1982 Falkland's War, a single Argentine Type 209 diesel submarine (ARA SAN LUIS) operated in the vicinity of the British task force for over a month. Despite the deployment of five nuclear attacked submarines, 24-hour per day airborne anti-submarine warfare (ASW) operations, and the expenditure of 203 British ASW weapons, the British task force never once detected the Argentine diesel submarine. The ARA SAN LUIS, on the other hand, had conducted several attacks on British ships but was unsuccessful because of an improperly maintained fire control system and poor quality torpedoes. What do you think the outcome of the war would have been if the ARA SAN LUIS had sunk or severely damaged one of the British small-deck carriers or logistics ships?

Air-Independent Propulsion Systems. Older diesel submarines must snorkel everyday or every other day to recharge their batteries. This evolution puts the diesel boat in a very vulnerable and detectable situation. To eliminate or reduce this vulnerability, air-independent propulsion (AIP) technology was developed. AIP operates without the need for outside air, is very quiet, and produces little heat. Third generation AIP systems allow diesel submarines to stay submerged for up to three months.

AIP technology does not have the international controls and restrictions that nuclear propulsion technology does. Therefore, countries do not have to get permission to develop and sell this technology. What are some of the countries that export AIP technology? Russia, Germany, Sweden, and France.

Mines. Like submarines, mines create the same psychological fear that comes from not knowing where they are located. They can limit the mobility of any aircraft carrier, surface combatant, amphibious ship, or submarine. It doesn't matter if they are simple contact and influence mines or a sophisticated one with a target detection system, mines are quite effective. In World War II, mines accounted for more ships damaged or lost than any other weapon. In Operation Desert Storm, enemy mines damaged the USS Tro[p;o (LPH 10) and USS Promcetpm (CG 59).

Nearly 30 nations manufacture mines; approximately 20 of these nations export their products. Currently, there are about 50 countries who maintain significant sea mine inventories (e.g., Iran, Iraq, North Korea, Cuba, Libya, Russia, and China). Mines can be deployed from almost any surface platform, including fishing boats, patrol craft, and merchant vessels. Mines can also be deployed by aircraft and submarines.

Submarine Significance

Submarines and anti-submarine warfare will remain the priorities of USW. Although U.S. Navy submarines have been around for 100 years, it wasn't until World War II that submarines finally demonstrated their significant contribution to naval superiority. German U-boats forced the Allies to commit disproportionately large forces to defend the Allied sea lines of communications. The Americans learned a lot from German U-boat design, construction, and tactics. We were able to significantly improve our submarines and torpedoes and successfully deploy them against the Japanese Navy. The results were impressive. U.S. submarines destroyed 1,314 enemy ships or 5.3 million tons. This equated to 60 percent of the Japanese tonnage lost from only 2 percent of our naval force. Their campaign was a critical factor in the industrial collapse of the Japanese war effort. But that success did not come without sacrifice. Of the 16,000 submariners, the Submarine Force lost 375 officers, 3,131 enlisted men, and 52 submarines.

Today, our nuclear submarines are vastly superior to their predecessors. Technological advances in all aspects of warfare have significantly improved submarine design, construction, sensors, communications, and weapons. These improvements only enhance the enduring characteristics of the submarine—stealth, endurance, firepower, mobility, and survivabil-ity. Thus, our submarines are multi-mission platforms that routinely conduct missions that no other platform can do.

Submarines excel at preparing and controlling the littoral battlespace for joint expeditionary forces. Based on numerous independent studies and intelligence assessments, 21st century regional powers are expected to have substantially improved capabilities to locate, target and engage non-stealthy platforms in the littorals. Submarines greatly enhance U.S. policymakers' understanding of enemy and terrorist force dispositions and operational doctrine before the outbreak of hostilities. Likewise they allow us to decisively engage and destroy key threats at minimal risk. Before a full aircraft carrier battlegroup or amphibious ready group with nearly 10,000 Sailors has approached a high threat area, a submarine can have already detected, reported and destroyed major threats.

The submarine force is charging ahead into the 21st century and is committed to enhancing the critical operational capabilities (Command, Control, and Surveillance, Battlespace Dominance, Power Projection, and Force Sustainment) of a modern naval expeditionary force.

Submarine Roles and Missions

The fundamental changes in the U.S. Submarine Force since the end of the Cold War involve major shifts in submarine warfighting concepts and doctrine, from the deterrence of global war to the support of U.S. national interests in regional crises and conflicts; from a primary Anti-Submarine Warfare (ASW) orientation against nuclear powered submarines to taking full advantage of the modern submarine's multi-mission capabilities; from weapon loadouts of primarily MK 48 torpedoes to Tomahawk Land-Attack missiles or other weapons. This changing operational context has rippled through all elements of U.S. submarine operations, from peace-time presence to strategic deterrence.

The transitions in the submarine force follow directly from the transitions in the world order and the evolving nature of the U.S. Navy. The world order has shifted from a bi-polar superpower alignment to a multi-polar collection of interests. While the likelihood of global conflict is greatly reduced, there is an increasing chance of regional conflict. The composition and operational posture of the U.S. Navy reflects this, having changed from a blue water emphasis to a littoral emphasis. For the submarine force this has meant several changes in roles:

- Prior to the end of the Cold War, Anti-Submarine Warfare was the major role for U.S. Attack Submarines. Now U.S. submarines are more multi-mission oriented.
- Intelligence gathering has shifted from strategic to tactical reconnaissance.
- The "Silent Service" is no longer completely silent, but exchanges information covertly with other U.S. forces.
- The submarine force is learning how to synergistically interoperate with other Navy and Joint communities for mutual mission accomplishment. This includes "community alliances" such as:
- Force Protection/Strike with Aircraft Carrier Battle Groups and Amphibious Ready Groups.
- Special Warfare with Special Operating Forces (such as Navy SEALs)
- Intelligence with the Surveillance community.

U.S. nuclear submarines perform numerous critical missions—many in ways that submarines are uniquely able to perform. To understand submarine tactics and strategy, one must first learn the capabilities of these ships. Many missions are classified; however, general mission areas include the following.

Intelligence, Surveillance and Reconnaissance (ISR)

Submarines provide the nation with a crucial intelligence gathering capability that cannot be replicated by other means. While satellites and aircraft are used to garner various types of information, their operations are inhibited by weather, cloud cover, and the locations of collection targets. In some situations it is difficult to keep a satellite or aircraft in a position to conduct sustained surveillance of a specific area. And, of course, satellites and aircraft are severely limited in their ability to observe or detect underwater activity.

Employed with multiple sensors and operated with care and cunning, submarines can monitor any event in the air, surface, or subsurface littoral domain providing a complete picture of an event across the full spectrum of intelligence disciplines. They are also an intelligence "force-multiplier" by providing tip-offs of high interest events to other collection assets. Submarines are able to monitor undersea events and phenomena not detectable by any other sensor. Since they are able to conduct extended operations in areas inaccessible to other platforms or systems, submarines can intercept signals of critical importance for monitoring international developments and enable a wide array of military operations. Furthermore, the ability to dwell covertly for extended periods defeats efforts to evade collection or deceive satellites and other sensors.

The unique look angle provided by a submarine operating in the littoral region enables it to intercept high interest signal formats that are inaccessible to reconnaissance satellites or other collection platforms. The intelligence gleaned from submarine operations ranges from highly technical details of military platforms, command and control infrastructure, weapons systems and sensors to unique intelligence of great importance to national policymakers on potential adversaries' strategic and operational intentions. Significantly, our submarines can provide real time alertment to National Command Authorities on indications of imminent hostilities. And unlike other intelligence collection systems such as satellites, submarines are also full-fledged warfighting platforms carrying militarily significant offensive firepower.

Precision Strike

U.S. attack submarines can carry 16 Tomahawk Land-Attack Missiles (TLAM) ready for submerged launch, with up to 12 additional Tomahawks that can be reloaded and fired while submerged. The TLAMs provide the capability for long-range, precision strike with conventional warheads against shore targets. Typically, submarines provide about 20 percent of the Tomahawk firepower in a carrier battle group. Additionally, because of their stealth, these attack submarines can be positioned to operate alone in environments where the risks would prevent surface and air forces from operating without extensive protective cover.

Other advantages of using submarines outfitted with Tomahawks are:

- Air superiority is not required
- Timely flexibility
- No chance of lost aircraft or airmen.

First used in combat in the 1991 Gulf War, the TLAM has proven to be a highly effective weapon. The official Department of Defense report "Conduct of the Persian Gulf War" (1992) states: "The observed accuracy of TLAM, for which unambiguous target imagery is available, met or exceeded the accuracy mission planners predicted." When the war began on the night of 16 January 1991, the opening shots were Tomahawk cruise missiles lunched from U.S. Navy surface ships in the Red Sea and the Persian Gulf. The missiles arrived over the heavily defended Iraqi capital of Baghdad at about the same time as U.S. Air Force F-117 "Stealth" attack planes carrying guided bombs. During the six week air war, F-117 attack planes were the only strike aircraft to operate over Baghdad at night, and TLAMs were the only U.S. weapons to strike the city in daylight during the entire campaign. Conventional aircraft were not used in strikes against Baghdad and certain other Iraqi targets because of the heavy anti-aircraft defenses. U.S. Navy surface ships and submarines fired 288 land-attack variants of the Tomahawk during the Gulf War. Battleships, cruisers, and destroyers launched 276 of the missiles and 12 were launched from submarines—the USS *Louisville* (SSN 724), operating in the Red Sea launched eight missiles and the USS *Pittsburgh* (SSN 720), operating in the eastern Mediterranean, launched four missiles. These launches demonstrated the ability of the submarine to operate as part of an integrated strike force, with targets and related strike data being communicated to them at sea. In future military operations submarines will not replace traditional carrier attack aircraft. Rather, submarine and surface ship-launched TLAM strikes will be the vanguard of such attacks, destroying early-warning, air-defense, and communications facilities to reduce the threats against manned aircraft. Submarines in particular can reach attack positions without alerting or provoking the intended adversary.

Special Operations Forces

Submarines have long been used for special operations—carrying commandos, reconnaissance teams, and agents on high-risk missions. Most special operations by U.S. submarines are carried out by SEALs, the Sea-Air-Land teams trained for missions behind enemy lines. These special forces can be inserted by fixed-wing aircraft, helicopter, parachute, or surface craft, but in most scenarios only submarines guarantee covert delivery. Once in the objective area, SEALs can carry out combat search-and-rescue operations, reconnaissance, sabotage, diversionary attacks, monitoring of enemy movements or communications, and a host of other clandestine and often high-risk missions. Nuclear-powered submarines are especially well-suited for this role because of their high speed, endurance and stealth. U.S. nuclear powered submarines have repeatedly demonstrated the ability to carry out special operations involving many swimmers. During exercises, which include Army, Air Force, and Marine Corps special operations personnel as well as SEALs, submarines recover personnel who parachute from fixed-wing aircraft and rappel down from helicopters into the sea, take them aboard, and subsequently launch them on missions. These Special Warfare Team Missions include:

- Combat Swimmer Attacks
- Reconnaissance and Surveillance
- Infiltration/Exfiltration Across the Beach
- Beach Feasibility Studies, Hydrographic Survey, and Surf Observation Teams in support of amphibious landing operations.

Any U.S. submarine can be employed to carry SEALs, however, the Navy has several submarines that have been specially modified to carry swimmers and their equipment more effectively, including the installation of chambers called Dry Deck Shelters (DDSs) to house Swimmer Delivery Vehicles (SDVs). These submarines retain their full suite of weapons and sensors for operations as attack submarines. But they have special fittings, modifications to their air systems and other features to enable them to carry Dry Deck Shelters. The DDS can be used to transport and launch an SDV or to "lock out" combat swimmers. A DDS can be installed in about 12 hours and is air-transportable, further increasing special operations flexibility. Several units of the STURGEON (SSN 637) class can carry one chamber each, while two former ballistic missile submarines can accommodate two shelters each. The DDS, fitted aft of the submarine's sail structure, is connected to the submarine's after hatch to permit free passage between the submarine and the DDS while the submarine is underwater and approaching the objective area. Then, with the submarine still submerged, the SEALs can exit the DDS and ascend to the surface, bringing with them equipment and rubber rafts, or they can mount an SDV and travel underwater several miles to their objective area. The number of SEALs carried in a submarine for a special operation varies with the mission, duration, target and other factors. One or more SEAL platoons of two officers and 14 enlisted men are normally embarked, plus additional SEALs to help with mission planning in the submarine and to handle equipment. Former SSBNs employed to operate with SEALs have special berthing spaces for about 50 SEALs.

Peacetime Engagement/Power Projection

In peacetime, the deployment of submarines in forward areas can demonstrate U.S. interest in the region. Alternatively, submarines are valuable if the President decides that interest should not be visible until a specific time. The long endurance and high transit speeds of nuclear submarines make them particularly attractive for rapid deployments to forward areas in such circumstances. Once on station the attack submarine can be highly visible—in 1991 U.S. submarines conducted more than 200 port visits to 50 cities around the world—or invisible. This operational flexibility is combined with the versatile firepower of the modern attack submarine. Submarines can also operate independently or in direct support of carrier battle groups, surface task forces, or with other submarines.

Whatever an opponent's ability to deny access to or preempt U.S. military presence, it can use these weapons in only limited ways against submarines. First, it cannot reliably detect their presence. Second, submarines are not threatened by many of the existing or projected access denial weapons. Coastal cruise missiles, tactical ballistic missiles and weapons of mass destruction pose little or no threat to a well-operated nuclear submarine. Submarines carry organic mine detection systems allowing them to avoid previously undetected minefields. A credible attack capability against our submarines could be developed only by substantial investment in an attack submarine force comparable to ours. Accordingly, so long as we maintain our investment advantage, submarines will remain one of the most credible, survivable and potent land attack missile platforms in our arsenal.

Sea Control

The United States is a maritime nation whose trade and military power projection capabilities depend upon assured use of the high seas. Ocean transport provides the vast majority—over 90 percent in most cases—of our strategic lift requirements. Keeping those sea lanes open and stopping enemy surface ships and submarines from using the seas is an important mission for submarines. Attack submarines can control the seas in a variety of scenarios, from general war against a major maritime power, to blockages of enemy ports. Attacks against enemy surface ships or submarines can be part of a war of attrition, where the object is to destroy as much of the opposing naval fleet or merchant shipping as possible, or such attacks can be directed against specific targets. Submarines are the quintessential sea control platforms, with proven anti-submarine and anti-surface capabilities. Several historical examples illustrate the power of submarines in naval warfare (e.g., WW II).

Modern U.S. submarines, armed with significantly improved sensors and weapons, are vastly superior to their historical ancestors. They possess unsurpassed abilities to hunt and kill submarines and surface ships on the high seas and in the littorals. U.S. nuclear submarines provide our only assured capability to wrest control of the sea from a determined enemy employing submarines in an area denial role. As a result, today's United States Navy, employing a combined arms anti-submarine capability that includes nuclear submarines, is able to sail freely on the world's oceans. And, as trade follows the flag, the merchant shipping of our nation, allies, and friends can conduct the trade on which our prosperity and security depend. Likewise, our power projection logistical military capability can be counted on to flow when and where needed.

Strategic Deterrence

Strategic deterrence remains a fundamental element of U.S. defense strategy, just as conventional deterrence has become increasingly important since the fall of the Berlin Wall. Nuclear-powered submarines will be the principal component of the future U.S. strategic posture. With the number of land-based bombers and intercontinental missiles being reduced; the SSBN force will be the only leg of the Strategic Triad still deploying missiles armed with Multiple Independently Targeted Reentry vehicles (MIRVs). The significance of the Navy's SSBN force was cited by General Colin Powell, U.S. Army, Chairman of the Joint Chiefs of Staff, at a ceremony in April 1992 marking the completion of the 3,000th deterrent patrol. General Powell told the submariners: "But no one—no one—has done more to prevent conflict, no one has made a greater sacrifice for the cause of Peace, than you. America's proud missile submarine family. You stand tall among all our heroes of the Cold War."

Because of the invulnerability of nuclear submarines operated in the vast ocean areas, they provide the nation's strategic deterrent more effectively and at less cost than other systems. Our TRIDENT submarines (SSBNs) now carry 54 percent of our nation's nuclear deterrent using less than 1.5 percent of naval personnel and 34 percent of our strategic budget. These Navy capital ships will form the backbone of the nation's strategic nuclear force well into the 21st century.

Task/Battle Group Operations

Attack submarines are fully integrated into Navy task/battle group operations. Typically, 2 attack submarines are assigned to each task group. These submarines participate with the battle group in all pre-deployment operational training and exercises. While operating with the battle group, tactical control of the submarines is routinely shifted to amphibious group commanders, battle group commanders, destroyer squadron commanders, or even NATO commanders. Likewise, tactical control of NATO submarines is routinely shifted to U.S. commanders.

Mine Warfare

In both covert offensive mining and mine reconnaissance, submarines provide capabilities that no other platform can deliver. The submarine offensive mining capability allows national leaders to precisely place mines for maximum effect without enemy alertment and with minimal risk. Mine reconnaissance capability from submarine launched Unmanned Undersea Vehicles allows the submarine to covertly detect and report mine danger areas without risk to naval forces. As a result, potential adversaries have fewer clues indicating potential locations of American expeditionary operations and U.S. military planners are better able to exploit the element of surprise.

Submarines carry mines to deny sea areas to enemy surface ships or submarines. Two types of mines are used by submarines, the enCAPsulated TORpedo (CAPTOR) and the Submarine- Launched Mobile Mine (SLMM). The CAPTOR can be used against submarines in deep water, while the SLMM is a torpedo-like weapon that, after being launched by the submarine, can travel several miles to a specific point, where it sinks to the sea floor and activates its mine sensors. It is particularly useful for blockading a harbor or a narrow sea passage.

Other USW Participants

As mentioned earlier, no one community can perform the Navy's undersea warfare role by itself. A combination of and integrated support between undersea, surface, airborne, and space-based systems are required to ensure that we maintain what the Joint Chiefs of Staff publication Joint Vision 2010 calls "full-dimensional protection." The undersea environment, ranging from the shallows of the littoral to the vast deeps of the great ocean basins—and polar regions under ice—demand a multi-disciplinary approach, subsuming intelligence, oceanography, surveillance and cueing, multiple sensors and sensor technologies, coordinated multi-platform operations, and underwater weapons. The following are short descriptions of other USW organizations rarely in the spot light, but whose missions and roles are becoming more critical to the Navy.

Integrated Undersea Surveillance System (IUSS)

The IUSS mission is to provide support for tactical and strategic forces through the detection, classification, tracking, and reporting of subsurface, surface, and air maritime activities. IUSS uses several systems and ships to accomplish their mission.

Sound Surveillance System (SOSUS) This fixed, sea floor acoustic system provides long-range detection of older and noisier classes of submarines. It has been installed for a number of years, and the SOSUS hydrophone arrays are positioned for the best acoustic intercept of contacts. The hydrophones are located throughout the Atlantic and Pacific Oceans and the Mediterranean and North Seas.

Fixed Distributive System (FDS) This is a long-life, low-frequency passive acoustic surveillance system for detecting new generation (quieter) submarines using hydrophones geographically distributed on the sea floor. It has been manufactured and installed within the last few years.

Other IUSS systems include the Surveillance Towed Array Sensory System (SURTASS), Advanced Deployable System (ADS), and the SURTASS Low Frequency Active (LFA) system. These systems are deployed on T-AGOS ships in the Atlantic and Pacific fleets.

Naval Meteorology and Oceanography Command (METOC)

The METOC mission is to provide the warfighter with the right METOC information, in the right format, to give him the decisive edge in combat. Interpreting the battlespace environment for our warfighters is the primary function of the Naval Oceanography community. Navy oceanographers are using every available resource to understand, predict, and portray the natural environment of waves, water, and weather, particularly in the littoral regions.

Other areas that the Naval Oceanography community is involved with are electronic intelligence and electro-optical collection, submarine launched mobile mines, shallow water ASW, ice cover, and bathymetry and digital nautical charts.

USW Platforms

Surface Ships

Aircraft Carriers: Aircraft carriers provide a wide range of possible response for the National Command Authority. Their mission is to provide a credible, sustainable, independent forward presence and conventional deterrence in peacetime; to operate as the cornerstone of joint/allied maritime expeditionary forces in times of crisis; and to operate and support aircraft attacks on enemies, protect friendly forces and engage in sustained independent operations in war. While aircraft carriers have no ASW sensors, they do carry the USW module for processing information from its aircraft and from shore facilities, as well as SH-60F S-3 aircraft.

TICONDEROGA Class (CG-47): These guided missile cruisers are large combat vessels with multiple target response capabilities. They perform primarily in a Battle Force role. These ships are multi-mission (AAW, ASW, ASUW) surface combatants capable of supporting carrier battle groups, amphibious forces, or of operating independently and as flagships of surface action groups. Due to their extensive combat capability, these ships have been designated as Battle Force Capable (BFC) units. The cruisers are equipped with Tomahawk ASM/LAM giving them additional long range strike mission capability. They carry MK 46 and MK 50 torpedoes and SH-60 and SH-3 helicopters.

ARLEIGH BURKE Class (DDG-51) and SPRUANCE Class (DD-963): These destroyers help safeguard larger ships in a fleet or battle group.

They operate in support of carrier battlegroups, surface action groups, amphibious groups and replenishment groups. Destroyers primarily perform anti-submarine warfare duty while guided missile destroyers are multi-mission (ASW, anti-air and anti-surface warfare) surface combatants. The addition of the Mk-41 Vertical Launch System or Tomahawk Armored Box Launchers (ABLs) to many *Spruance*-class destroyers has greatly expanded the role of the destroyer in strike warfare. They carry MK 46 and MK 50 torpedoes and SH-60 and SH-3 helicopters.

OLIVER HAZARD PERRY Class (FFG-7): Guided missile frigates fulfill a Protection of Shipping (POS) mission as ASW combatants for amphibious expeditionary forces, underway replenishment groups and merchant convoys. FFG's bring an anti-air warfare (AAW) capability to the frigate mission, but they have some limitations. Designed as cost efficient surface combatants, they lack the multi-mission capability necessary for modern surface combatants faced with multiple, high-technology threats. They carry MK 46 and MK 50 torpedoes and SH-60 helicopters.

FFG's have a limited capacity for growth. Despite this, the FFG-7 class is a robust platform, capable of withstanding considerable damage. This "toughness" was aptly demonstrated when USS *Samuel B. Roberts* struck a mine and USS *Stark* was hit by two Exocet cruise missiles. In both cases the ships survived, were repaired and returned to the fleet. USS *Stark* was recently decommissioned in May 1999. The Surface Combatant Force Requirement Study does not define any need for a single mission ship such as the frigate and there are no frigates planned in the Navy's five-year shipbuilding plan.

Aircraft

P-3 Orion: Fixed-wing shore-based maritime patrol aircraft (MPA). Carries radar, MAD, FLIR, sonobuoys, EA, and an onboard computer for processing information from sonobuoys. Weaponry includes Harpoon, torpedoes, and mines.

SH-60B LAMPS III: Helicopter developed to operate from cruisers, destroyers, and frigates. Sensors include radar, MAD, EA, sonobuoys, and FLIR. The SH-60B can also carry two MK 46 torpedoes. Sonobuoys are processed on the helo or linked back to and processed on the surface platform. The data link is a direct pencil beam in the SHF frequency range which is not likely to be detected by enemy ES. The link also contains voice comms, nav data, and ES. When not operating in the USW mode the SH-60B provides excellent ASUW support by linking its radar data to the surface platform.

SH-60F: This carrier-based variant of the LAMPS III is similar in most respects to the Bravo version with the addition of a dipping sonar. Because it has been adapted for close-in USW defense of the aircraft carrier, the radar, MAD, and sonobuoys have been deleted.

Submarines

LOS ANGELES Class (SSN-688): This is the most numerous class of nuclear-powered fast attack submarines built by any nation, and will form the backbone of the U.S. attack submarine force well into the 21st century. LOS ANGELES Class submarines are fast and can carry 25 torpedo-tube launched weapons. The last 31 hulls of the class have 12 vertical launch tubes for the Tomahawk cruise missile. Of these, the final 23 hulls, referred to as "688I" (for improved), are quieter, incorporate an advanced combat system, and are configured for under-ice operations. (Their forward diving planes were moved from the sail structure to the bow and the sail has been strengthened for breaking through ice.) The USS Memphis (SSN 691) was modified to serve as a test and evaluation platform for advanced submarine systems and equipment, while retaining her combat capability.

SEAWOLF Class (SSN-21): The successor to the 688-class, this submarine has improved machinery, quieter propulsion, and improved weapons systems. The SEAWOLF is significantly quieter and faster than the L.A. class and also has more torpedo tubes. Originally intended as a class of 29 submarines to be built during a ten-year period, the end of the Cold War and budget constraints led to a revision in submarine planning. Only three submarines of this design have been built. The SSN 21 can carry more weapons—up to 50 torpedoes or missiles, or up to 100 mines—than any other U.S. submarine. The SEAWOLF also introduces the advanced AN/BSY-2 combat system, which includes a new larger spherical sonar array, a Wide Aperture Array (WAA), and a new towed-array sonar.

VIRGINIA Class (SSN-774): Designed for multi-mission operations and enhanced operational flexibility. SEAWOLF-class quieting has been incorporated in a smaller hull while military performance has been maintained or improved. With a focus on the littoral battlespace, the VIR-GINIA has improved magnetic stealth, sophisticated surveillance capabilities, and Special Warfare enhancements.

VIRGINIA is engineered for maximum design flexibility. Its responsiveness to changing missions and threats, and the affordable insertion of new technologies, ensures it will continue to be the right submarine well into the 21st Century. Integrated electronic systems with commercial-off-the-shelf (COTS) components facilitate state-of-the-art technology introduction throughout the life of the class and avoid unit obsolescence.

The Command, Control, Communications, and Intelligence (C3I) electronics packages also promote maximum flexibility for growth and upgrade. Coupled with the Modular Isolated Deck Structure (MIDS) and open-system architecture, this approach results in a significantly lower cost, yet more effective, command and control structure for fire control, navigation, electronic warfare, and communications connectivity.

The VIRGINIA's sonar system is state-of-the-art and has more processing power than today's entire submarine fleet combined to process and distribute data received from its spherical bow array, high-frequency array suite, dual towed arrays, and flank array suite.

The VIRGINIA's sail configuration houses two new photonics masts for improved imaging functions, and improved electronics support measures mast, and multi-mission masts that cover the frequency domain for full-spectrum, high data-rate communications. The sail is also designed for future installation of a special mission-configurable mast for enhanced flexibility and warfighting performance.

VIRGINIA is armed with a variety of weapons. It carries the most advanced heavyweight torpedoes, mines, Tomahawk cruise missiles, and Unmanned Undersea Vehicles (UUVs) for horizontal launch. In addition, Tomahawk missiles are carried in vertical launch tubes. The submarine also features an integral Lock-Out/Lock-In chamber for special operations and can host Special Operations Forces' underwater delivery vehicles.

Reducing acquisition and life-cycle costs is a major objective of the VIRGINIA-Class submarine design and engineering process. Substantial cost avoidance is being achieved through the application of concurrent engineering design/build teams, computer-aided design and electronic visualization tools, system simplification, parts standardization, and component elimination. These innovations ensure that the ship is affordable in sufficient numbers to satisfy America's future nuclear attack submarine force level requirements.

The VIRGINIA Class Program Office is applying the lessons learrned from successful government and industry programs of similar scope and complexity to improve production and lower costs. Integrated Product and Process Development (IPPD) teams bring the combined experience of the shipbuilders, vendors, designers and engineers, and ship operators to bear on the ship design. The early involvement of production people on these teams ensures an excellent match between the design and the shipbuilder's construction processes and facilities, allows a smoother transition from design to production, and reduces the number of changes during construction. The ship is designed using a state-of-the-art digital database, which allows all members of the IPPD teams to work from a single design database and provides three-dimensional electronic mockups throughout the design process. These efforts, along with strong support from Navy and shipbuilder management, result in an affordable submarine that satisfies all operational needs. The USS *Virginia's* keel was laid in September 1999 and is scheduled for delivery in 2004.

The military performance of the VIRGINIA Class submarine is comparable to that of the SEAWOLF Class, with significant improvement in littoral warfare capabilities and considerably less cost. The New Attack Submarine will be stealthier than SEAWOLF. It surpasses the performance of any current projected threat submarine, thus ensuring U.S. undersea dominance well into the next century.

SSBN 726, OHIO Class: The OHIO Class fleet ballistic missile submarines provide the sea-based leg of the Triad of U.S. strategic forces with the 18 Trident SSBNs each carrying 24 missiles. By virtue of their patrol posture, these submarines are highly survivable; they are also extremely flexible, capable of rapidly retargeting their missiles, should the need arise, using secure and constant at-sea communications links. They are the largest submarines to be built by the United States. The USS *Ohio* made the first operational patrol of this class in the fall of 1982. The 18th and final ship of this class was delivered to the Navy in 1997. OHIO Class submarines can carry either the TRIDENT I (C-4) or TRIDENT II (D-5) missiles. In addition, these submarines are fitted with four torpedo tubes for MK 48 torpedoes which, along with countermeasure devices, provide defense against hostile ASW forces (offense if all missiles away!) The most important defensive feature of these submarines is their stealth—they are among the quietest nuclear-powered submarines ever built. This inherent feature of the OHIO Class coupled with other characteristics makes these ships the most survivable element of the nuclear Triad.

Two complete crews—designated Blue and Gold—are assigned to each OHIO Class submarine. While one crew is at sea operating the submarine, the other is conducting training, attending schools, being evaluated in shore-based simulators, and enjoying leave. By alternating the Blue and Gold crews, with a brief turnover period, the submarines can be kept at sea for considerably longer than with a single crew. The nominal operating schedule is 77 days at sea followed by a 35 day turnover/replenishment/refit period. OHIO Class submarines are specifically designed for extended deterrent patrols. To speed the time in port for crew turnover and replenishment, three large logistics hatches are fitted to provide large diameter resupply and repair openings. These allow the rapid transfer of supply pallets, equipment replacement modules, and even machinery components, significantly reducing the time required for replenishment and maintenance. The OHIO design and modern maintenance concepts permit the submarines to operate for over 12 years between major shipyard availabilities.

Platform Strength and Weaknesses

PLATFORM	STRENGTHS	WEAKNESSES
SUBMARINE	- Quiet - Good weapon—MK 48 - Long on-station time - Relatively large weapon load - Good acoustic sensors - In the medium	- Communications - Order modifications (slow to respond to changes) - Operate slowly for maximum range detections - Tough to use ES, radar
SURFACE	- Low noise (with Prairie Masker) - Large Weapon/sensor load - C3 Facilities - Organic Helo - Additional sensors - Weapon reach - Long on-station time - Can operate at all depths with sensor placement	- Short punch without a helo - Vulnerable to torpedo attack - Vulnerable during Unrep - Weapon (MK 46) has questionable kill capability
AIR	- Speed - Good radar/ES coverage - Helo with dipper - Extends ships sensors/weapons reach - Not vulnerable to attack - Optimum placement	- Low on-station time (P-3 max about 12 Hrs.) - Small weapons load - No UNREP capability - Communications vital for entire mission - Station keeping/navigation may be difficult

USW Weapons

Torpedoes

Self-propelled guided projectiles that operate underwater and designed to detonate on contact or in proximity to a target. Torpedoes may be launched from submarines, surface ships, helicopters and fixed-wing aircraft. They are also used as parts of other weapons; the Mark 46 torpedo becomes the warhead section of the ASROC (Anti-Submarine ROCket) and the Captor mine uses a submerged sensor platform that releases a torpedo when a hostile contact is detected. The three major torpedoes in the Navy inventory are the Mark 48 heavyweight torpedo, the Mark 46 lightweight and the Mark 50 advanced lightweight.

The MK-48 is designed to combat fast, deep-diving nuclear submarines and high performance surface ships. It is carried by all Navy submarines. The improved version, MK-48 Advanced Capability (ADCAP) is replacing the MK-48s. The MK-48 has been operational in the U.S. Navy since 1972. MK-48 ADCAP became operational in 1988 and was approved for full production in 1989. MK-48 and MK-48 ADCAP torpedoes can operate with or without wire guidance and use active and/or passive homing. When launched they execute programmed target search, acquisition and attack procedures. Both can conduct multiple reattacks if they miss the target. The MK-48 has a range of greater than 5 miles, can dive greater than 1,200 ft, can travel greater than 28 knots, and carries 650 lbs of high explosive.

The MK-46 torpedo is designed to attack high performance submarines, and is presently identified as the NATO standard. The MK-46 Mod 5 torpedo is the backbone of the Navy's lightweight ASW torpedo inventory and is expected to remain in service until the year 2015. The MK-46 torpedo is designed to be launched from surface combatant torpedo tubes, ASROC missiles and fixed and rotary wing aircraft. In 1989, a major upgrade program began to enhance the performance of the MK-46 Mod 5 in shallow water. Weapons incorporating these improve-ments are identified as Mod 5A and Mod 5A(S). The MK-46 has a range of 4 miles, can dive greater than 1,200 ft, can travel greater than 28 knots, and carries 98 lbs. of PBXN-103 high explosive (bulk charge).

The MK-50 is an advanced lightweight torpedo for use against the faster, deeper-diving and more sophisticated submarines. The MK-50 can be launched from all ASW aircraft, and from torpedo tubes aboard surface combatant ships. The MK-50 will eventually replace the MK-46 as the fleet's lightweight torpedo. The MK-50 has a range of greater than 4 miles, can travel greater than 40+ knots, and carries approximately 100 pounds of high explosive (shaped charge).

Mines

Mines are used as an anti-ship or anti-submarine, subsurface weapon. The MK56 ASW mine (the oldest still in use) was developed in 1966. Since that time, more advances in technology have given way to the development of the MK60 CAPTOR (short for "encapsulated torpedo"), the MK62 and MK63 Quickstrike and the MK67 SLMM (Submarine Launched Mobile Mine). Most mines in today's arsenal are aircraft delivered to target. Mines will be covered in the mine warfare section.

Submarine Task Group Operations

Attack submarines assigned to task groups is a relatively new way of doing business in the Navy. Submarines and task group commanders have had to learn how to operate together and learn each other's way of doing business. The benefits for the Officer in Tactical Command (OTC) are tremendous. The OTC now has another platform at his disposal and can use the submarine in a variety of roles (e.g. providing intelligence data or clearing the battle space prior to the task force's arrival).

Terms

Coordinated submarine/task group operations require a strong command and control (C^2) arrangement to avoid conflicting tasking requirements from different commanders for the same platform and to ensure safety of the submarine. To understand the C^2 structure, the student must understand the terms associated with submarine/task group operations.

The submarine operating authority (SUBOPAUTH) is the naval commander exercising operational control of submarines. The submarine force commander or his designated subordinate for a specified area acts as SUBOPAUTH for overall submarine operations.

The officer in tactical command (OTC) is the senior officer present eligible to assume command or the officer to whom he has delegated tactical command. The OTC is usually the commander of a task force or group.

Operational command is the authority granted to a commander to assign missions or tasks to subordinate commanders, to deploy units, to

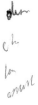

reassign forces, and retain or delegate operational or tactical control as necessary. The equivalent of this C^2 term in the U.S. joint arena is COCOM. It is a nontransferable function performed at the area command level by a unified or specified commander and carries with it the authority to delegate operational control of forces to subordinate commander(s).

Operational control (OPCON) is the authority delegated to a commander to direct forces assigned so the commander may accomplish specific missions or tasks which are usually limited by function, time, or location; to deploy units concerned; and to retain or assign tactical command (TACOM) or control of those units. In naval or joint operations, SUBOPAUTH is designated to exercise OPCON over all assigned submarines. This includes responsibility for overall area-wide water space management (WSM) and prevention of mutual interference (PMI) for submarine operations with the commander-in-chief's or COCOM's area of responsibility (AOR). SUBOPAUTH also has responsibility for and control of the submarine broadcast.

Tactical command (TACOM) is the authority delegated to a commander to assign tasks to forces under his command for the accomplishment of the mission assigned by higher authority. TACOM includes retention of authority to delegate tactical control (TACON). TACOM of submarines assigned to a joint or naval task force/group can be delegated to an at-sea commander as mutually agreed by the OTC and SUBOPAUTH.

Tactical control is the detailed and usually local direction and control of movements or maneuvers necessary to accomplish missions or tasks assigned. The officer exercising TACON also acts as the weapons control authority for units assigned as consistent with the rules of engagement (ROE). For submarine operations, TACON is confined to the area assigned to each individual submarine, either by the SUBOPAUTH or the OTC, if TACOM was shifted. The submarine is still required to copy the submarine broadcast as directed by the SUBOPAUTH regardless of the communications and reporting requirements levied by the officer exercising TACON.

Waterspace management (WSM) pertains to the allocation of waterspace in terms of anti-submarine (ASW) weapons control to permit the rapid and effective engagement of enemy submarines, while preventing inadvertent attacks on friendly submarines. Essentially, WSM is a set of specifically defined submarine and ASW force operating areas and attack rules. These procedures are implemented by the SUBOPAUTH on behalf of the area or joint operations combatant commander, and should be in place whenever use of ASW weapons by any platform becomes probable.

WSM may be applied on any scale—local, regional, theaterwide—depending on the crisis situation, the existing or projected submarine threat, and requirements of the area or combatant commander.

Prevention of Mutual Interference (PMI) procedures are specifically intended to prevent submerged collisions:

1. Between friendly submarines
2. Between submarines and friendly surface ships' towed bodies
3. Between submarines and any other underwater event (e.g., explosive detonations, research submersible operations, oil drillings, etc.).

C² of Submarines Employed with Surface Forces

Submarines are assigned to tasks groups at one of the four levels of command and control and coordination: Integrated Operations, Direct Support, Associated Support, and Area Operations. The SUBOPAUTH and the OTC should agree upon the level of command and control desired based on mission, C² capabilities of units assigned, and operational responsiveness required. TACOM of submarines and responsibility for local WSM and PMI can be delegated to the OTC. But in all cases, SUBOPAUTH retains OPCON of all assigned submarines.

Integrated Operations allows the OTC to exercise TACOM of assigned submarines. OTC assumes the responsibility for all operations and safety of assigned submarines, including local WSM and PMI for their designated area.

In Direct Support operations, the SUBOPAUTH retains TACOM and shifts TACON to an afloat commander. The OTC or designated subordinate commander directly controls the tactical movement and actions of assigned submarines in specific waterspace areas designated by the SUBOPAUTH.

In Associated Support operations, submarines operate under the TACON of the SUBOPAUTH, who coordinates specific unit tasking and movement in response to requirements of the OTC or designated subordinate commander. In this role, the submarine communicates with the supported force for exchange of intelligence and is tasked to coordinate operations with elements or units of that force.

In Area Operations, submarines are operating independently of the task force but may be specifically tasked by the SUBOPAUTH in roles that support the particular objectives of a surface force. These tasks are normally executed autonomously with no requirement for the submarine to communicate or cooperate with the supported force. The SUBOPAUTH informs the task force/group commander concerning status and completion of submarine tasks assigned.

Submarine Search and Attack

Stealth is one of the submarine's fundamental characteristics that make the submarine such a formidable platform. Unfortunately, the "silent service" cannot reveal in more detail how a submarine operates without getting into higher levels of classification. RADM Fages, Director, Submarine Warfare Division (N87), wrote, "Stealth allows the imposition of force at the time and place of one's choosing. It creates uncertainty in the mind of an adversary, and it imposes financial and operational costs to counter the submarine that may be lurking in his littorals."

Attack Philosophy

The successful commanding officer (CO) will maintain stealth and the element of surprise until he is ready for the kill. Once his weapons are away, the submarine CO knows that his presence has been compromised. Therefore, the CO will delay alerting his enemy of his presence for as long as possible. When hunting enemy submarines or other warships, the submarine commanding officer will accept risks and reduce the margin of safety normally maintained in peacetime.

To maintain a high state of readiness, submarines constantly train. Be it with ship drills, watch section training, or divisional training-the ship constantly practices for the real thing. In addition to the training, the submarine also maintains peak system performance. This is accomplished through weapons systems alignment checks, fire control and sensor preventive maintenance, and self radiated noise monitoring.

There are three phases the submarine follows when destroying an enemy warship:

Contact Phase

This is the period of time from initial detection of a contact until determination of tentative classification and initial range assessment. Detection of a contact is accomplished by active or passive sonar or by visual means. All initial contacts are treated as potential threats unless proven otherwise.

Approach Phase

When the contact is classified as a threat, the fire control team will develop a target motion solution to obtain adequate firing information. The submarine may close the contact as necessary to get better information.

Attack Phase

This phase begins when the CO decides to put some ordnance in the water and ends when the target is sunk or when the enemy disengages.

Task Force USW Operations

The *USWC (Undersea Warfare Commander)* may be a Commanding Officer of a ship, or embarked staff (ex: COMDESRON 12). Reports to CWC all matters concerning USW operations. The USWC is in charge of all weapons and sensors on all USW ships in the Battle Group.

A *Search and Attack Unit* (SAU) is one or more ships separately organized or detached from a formation as a tactical unit to search for and destroy submarines. The SAU may be augmented by fixed-wing aircraft or helicopters. The SAU commander is responsible for:

(1) *Deciding best approach to datum* (last known position of sub).
(2) *Promulgating search and attack plans* (from ATP 1(C)—Vol I).
(3) *Designating attacking unit.*
(4) *Keeping USWC informed.*

There are several factors to consider when organizing a task force that will operate in an USW environment, among them:

(1) *Mission.*
(2) *Threat* - Diesel or nuclear sub? If threat is diesel subs task force will normally use active sonar (passive for nuclear). What weapons do they carry?
(3) *Capabilities and limitations of own forces.*
(4) *Area of operations*—Open ocean, choke points.

Undersea Warfare operations are classified as *protective* and *offensive actions*. USW protective actions include escort/screen duties, support operations, or harbor defense. In offensive USW, USW units search out, locate, and attack enemy submarines. Protective or defensive USW normally occurs when the presence of an enemy submarine is unknown. When that presence becomes known, or when an enemy sub is spotted, offensive USW is used to attack or shoulder.

A submarine is said to be operating within a specific zone relative to your position or the position of the high value unit. The Outer, Middle and Inner Zones make up the USW geographic area. Objectives and Platforms associated with these zones are tabulated in the following table:

ZONE	OBJECTIVES	PLATFORMS
OUTER (Normally 100–300 nm from main body but can extend out to 1000 nm)	- Detect subsurface contacts, in enough time to be able to divert Battle group around the threat	- MPA - SSN in direct support - Tail ships
MIDDLE (30–100 nm from	- Sanitize base course main body) - Localize, track, and classify all contacts - If ROE permits attack	- Tail ship - P-3, S-3, SH-60, SH-3
INNER (0–30 nm from main body)	- Same as Middle Zone, except the time for engagement is shorter. Attack from enemy torpedoes is considered imminent. Maximum self defense measures must be implemented.	- Main body units (HVU) and escorts - All air except P-3 and S-3 - Tail and Hull mounted sonar surface assets

Protective USW Actions

The purpose of protective USW actions is to protect the main body and or a convoy. A technique was developed to aid the USWC in determining danger positions by plotting the submarine weapons capability and intercept problem.

Torpedo Danger Zone (TDZ) This is a 10,000 yd. circle plotted around the main body or convoy, taking the formation speed into consideration, that a submarine must enter to be within effective conventional torpedo range. The faster the formation speed, the more skewed the TDZ is plotted ahead.

Limiting Lines of Approach (LLOA). The limiting lines of approach determine whether a submarine, knowing the course and speed of a force, can reach the TDZ in a direct line. These lines are plotted tangent to the TDZ circle. As the formation speed increases, the LLOA will tend to narrow. Conversely, as the formation speed decreases the LLOA tend to widen making USW defense more difficult.

Zig-Zag Patterns. Often formations and convoys will alter course at reqular intervals around a base course. This zig-zagging makes it harder for a submarine to obtain a fire control solution.

Offensive USW Actions

Offensive USW actions attempt to neutralize the sub threat when the enemy sub presence is identified.

Datum. The last known position of a submarine. This position is plotted and is the starting point of a prosecution. Obviously, time is of the essence. Navigation and sensor inaccuracies must be considered when plotting Datum Error. This is the area of ocean that an enemy sub might actually be operating.

Furthest on Circles (FOC). These are circles that expand from datum based on time and estimated speed of the sub. It is very similar to plotting a D.R. only this is 360 degrees around datum.

Torpedo Danger Area (TDA). This is approximately a 8,000 yard extension plotted beyond the most recent FOC, where an entering prosecuting ship must take due precautions to avoid a torpedo attack.

Cone of Courses (COC). The range of course and speed combinations that a sub may use to intercept the main body/convoy TDZ from datum. This provides the initial search area for the prosecuting assets en route in an offensive role.

Approaches To Datum. There are several different techniques used by prosecuting assets when approaching datum in relation to the position of the main body/convoy.

(1) *DIRECT APPROACH*: Used by attacking assets when the position of datum (or sub) is within 6 nm beyond TDZ.

(2) *OFFSET APPROACH*: Used by attacking assets when the distance to datum (or sub) is greater than the 6 nm beyond TDZ but within +/- 30 degrees of main body/convoy track. The offset is 10 to 30 degrees from a direct course to datum offset towards the formation course.

(3) *INTERCEPT APPROACH*: Used by attacking assets when the distance to datum (or sub) is greater than the 6 nm beyond TDZ and also greater than 30 degrees of the main body/convoy track. A MoBoard intercept solution is used for the approach.

Search and Attack Method In the past, this was predominantly a ship function, with aircraft in an assist role. With the increase in capability in passive detection (TACTASS, sonobuoys) the roles have almost reversed, with the intent to detect, localize, classify, track and attack well beyond the weapons range of the threat submarine or the searching ship.

STEP 1: *DETECTION*—May come first in the form of a passive detection on a surface ship's Tactical Towed Array Sonar System (TACTASS). First detection may also come from a sonobuoy field laid by an S-3 or P-3C. Seldom, the initial detection is from a helo. Helicopters are typically response platforms and not search platforms due to the low number of sonobuoys onboard, and relative short on-station time.

STEP 2: *CLASSIFICATION* - Classification is the decision as to whether a contact is or is not a submarine. There are four classifications.

(1) *CERTSUB*—A contact that has been sighted and positively identified as a submarine is classified CERTSUB (certain submarine) or a weapon is fired.

(2) *PROBSUB*—A contact that displays strong cumulative evidence of being a submarine is classified PROBSUB (probable submarine). This classification is based on the evaluation of data from one or more sensors (radar, sonar, MAD, sonobuoy, etc).

(3) *POSSUB*—The classification POSSUB (possible submarine) is applied to a contact on which available information indicates the likely presence of a submarine, but on which there is insufficient evidence to justify a higher classification. The classification POSSUB must always be accompanied by an assessment of the confidence level as follows:

Low confidence—A contact that cannot be regarded as a non-submarine and which requires further investigation.

High confidence—A contact which, from the evidence available, is firmly believed to be a submarine but does not meet the criteria for PROBSUB. A single sensor contact usually meets this criteria.

(4) *NONSUB*—The evaluator is entirely satisfied that the contact is not a submarine.

STEP 3: *LOCALIZATION*—Is the process in which the contact's location is better defined. Because TACTAS is streamed in a line behind the surface platform, the initial passive detection can only be described as a relative bearing from the "tail." Consequently, there are two solutions for each relative bearing. This is known as bearing ambiguity, which can be resolved either by geography, intelligence, other sensors, or Target Motion Analysis (TMA). The TMA method to resolve bearing ambiguity requires turning the ship to a new heading and finding the new line of relative bearing on the tail. The bearing ambiguity is then resolved, because the new relative bearing will intersect only one of the first two bearings.

Once bearing ambiguity has been resolved, localization can be further accomplished by sending a helo down the line of bearing with sonobuoys and MAD. Once contact is made with the second sensor a sonobuoy field can be laid to better refine the position.

In addition, TMA from the MK116 underwater fire control system (UWFCS) can be based on MLE (Maximum Likelihood Estimator) automatically or by MATE (Manual Adaptive TMA Evaluator).

STEP 4: *TRACKING*—The contact is then tracked using either active or passive means. If possible, passive tracking is the first choice so as to not alert the target that he is being tracked. There are pros/cons of active and passive sensors, and their use against nuclear and diesel targets. The objective in tracking is to determine course, speed, depth and to obtain a targeting solution or attack criteria.

STEP 5: *ATTACK*—The placement of a weapon on the target. Only one at a time, torpedoes can interfere with each other. Once a torpedo is in the water you should also start active tracking if not already. The platform of choice is an aircraft so as not to place a surface ship or submarine within the enemy's weapons range.

(1) *URGENT ATTACK*: An attack conducted urgently with less regard to exact submarine location and high emphasis on timeliness. Even without exact location of the submarine, placing a torpedo any-

where in the general area of a submarine gives him a problem that must be dealt with. He now has to worry more about defending himself and less about sinking the target. If nothing else, it tends to buy time.

(2) *DELIBERATE ATTACK*: As the name implies, precise location and accuracy are of essence rather than timeliness. This tends to occur when then main body is well protected or has been diverted and is in no immediate danger from a torpedo.

Review Questions

1. What is USW?
2. Why is USW important?
3. What is asymmetric warfare?
4. How does asymmetric warfare affect modern warfare?
5. Why are diesel submarines a threat to a carrier battle group?
6. How are modern submarines different from their predecessors?
7. Identify how submarine roles and missions have changed since the end of the Cold War. Why?
8. What are some strengths and weaknesses of the various USW platforms?
9. What is waterspace management?
10. What is the difference between SUBOPAUTH and OTC?
11. What is the SAU commander responsible for?
12. Identify and describe two techniques used by surface combatants to hinder a submarine's ability to obtain a firing solution?
13. What is meant by protective and offensive USW measures?
14. What are some factors and considerations for the Task Force in USW?
15. What are the three USW geographic divisions in USW?
16. What are the classification criteria used for submerged contacts?

7

Air Warfare

Learning Objectives

At the end of this lecture the student will be able to:

Comprehend how Air Warfare doctrines contribute to the basic sea control and power projection missions of the naval service. More specifically:

Define AW.

Describe offensive and defensive measures of AW.

Describe the concept of Defense in Depth.

Describe the phases of AW.

Describe the various platforms and weapons involved in AW.

Describe the factors and considerations that affect AW planning.

Additional Required Reading

None.

Air Warfare (AW) is defined as the actions required to destroy or reduce the enemy air and missile threat to an acceptable level. It includes both offensive and defensive measures.

(1) *Offensive measures* include strikes against ships, air bases and missile sites. Often these offensive measures fall under the

purview of the Strike Warfare Commander. The AWC and STWC must therefore work together closely under the direction of the CWC.

Offensive measures are covered in some detail in the chapter on "Strike Warfare."

(2) *Defensive measures* include the use of interceptors, surface to air missiles (SAM) and air to air (AAM) missiles, guns, electronic countermeasures, cover, concealment, dispersion, and mobility to protect or defend against an offensive threat.

Successful conduct of defensive AW involves the proper employment of the *Defense in Depth* concept. This concept includes the coordination of detection equipment, weapons systems, and communications and depends on the tactics employed, the disposition of the force, the operating efficiency of equipment and the capability of personnel.

Defense in Depth

The Defense in Depth concept includes an Air Warfare area encompassing the total region to be protected from enemy air attack. The area is divided into three areas: *Surveillance Area, Destruction Area and Vital Area.*

Surveillance Area

The surveillance area extends from the center of the area to be protected to the maximum detection range of the battle group. This detection range depends upon the location of the ships and aircraft in company and the type of sensors being employed. For this reason the surveillance area should not be thought of as a circle but rather an area that is constantly changing as units move around using different sensors.

Classification, Identification, and Engagement Area (CIEA)

The CIEA is that region of coverage in which AW weapons may be employed against an air threat. The actual size and shape of this area depends on the geographical position of different AW assets and their own weapons capabilities. It may be further divided into zones:

- Fighter Engagement Zone (FEZ). Engage enemy with air assets.
- Missile Engagement Zone (MEZ). Engage enemy with SAM.

FA-18D Hornet

- Close-in Engagement Zone (CEZ). Engage enemy with guns, CIWS, and BPDMS.
- Joint Engagement Zone (JEZ). Engage enemy with multiple air defense weapon systems (SAM and fighters) of one or more Service components simultaneously. Successful JEZ operations require that all air defense systems are capable of discerning between enemy, neutral, and friendly air vehicles in a highly complex environment. If this requirement can not be met, separate zones for fighter and missile engagement zones should be established.

In the CIEA, the AWC must identify all air contacts and classify them as friendly, unknown, or hostile. There are a variety of methods for identifying air contacts; it is vital that AW watchstanders be familiar with all of them and be able to use the data they learn to make instantaneous decisions whether to engage or not engage a target. No single ID method can provide a positive classification and identification; they must be used together to help the watchstander make an analysis of his target. These methods include:

- Identification Friend of For (IFF) is the use of radar transponder codes to determine whether or not an aircraft is friendly. There are four modes of IFF; the first three are used by civilian and military aircraft to identify themselves to air traffic controllers. All U.S. and NATO aircraft are capable of transmitting *Mode 4* signals, which are encrypted and identify the aircraft as a definite "friend."
- Profile—by examining an aircraft's flight profile, a radar operator can determine its speed, position, direction of flight, point of origin, and (depending on the type of radar) its altitude. This information can be used to make an educated guess about a type of aircraft. For instance, an aircraft flying at 35,000 feet, 500 knots, on a commercial air route appears to be a commercial airliner. However, an aircraft travelling at Mach 2 is almost certainly a military aircraft.
- Indications & Warning (I & W)—Intelligence sources can provide information about a specific aircraft or a possible threat than can correlate with a newly detected aircraft.
- Electronic Sensing (ES)—ES equipment can detect and classify emitters from unknown aircraft of missiles and provide analysis that helps to identify the target.
- Visual Identification (VID)—personnel on ships or in a friendly airplane can identify unknown airplanes/missiles by sight and pass that information to AW.

Vital Area

The Vital area contains the unit or units crucial to mission success and is considered to extend from those units to the maximum weapons release range of the enemy's weapons.

Phases of Air Warfare

There are two major phases of an AW operation: *Surveillance/Detection* and *Engagement*.

Phase 1: Surveillance/Detection

Phase 1 is a continuous operation. At no point is the intensity of phase one lowered or ceased, despite the threat situation. Working forward, from the sea in the littoral environment, time is of the essence. Low, slow flyers can be the greatest challenge to our sensors. Working close to land it is imperative to identify an unknown air contact and have a fire control solution prior to the contact reaching weapons release range. *It is much easier to shoot the archer than the arrow.*

AIRBORNE EARLY WARNING AIRCRAFT
Timely warning of airborne threats is enhanced by employing *airborne early warning (AEW) aircraft.* AEW aircraft are equipped with an airborne tactical data system (ATDS) that is capable of linking with the shipboard NTDS. They are very good at detecting low-flying aircraft and missiles, can conduct air controlled intercepts, and act as a communication relay between the force and distant stations.

PICKET UNITS
Picket units, which are stationed in an air corridor through which all friendly aircraft must pass when returning so that they may be "deloused" of any shadowing enemy aircraft and given safe passage to recovery. Any combatant ship or AEW aircraft capable of air intercept control may be assigned picket duties in addition to their surveillance function. AW pickets are explained in greater depth below:

(1) *RADCAP.* Fighter aircraft assigned as airborne radar picket.
(2) *PIRAZ/SSS (positive identification radar advisory zone/ strike support ship)* is an NTDS equipped missile ship that can simultaneously track, analyze, and display multiple air contacts and broadcast them over link 11. Ships assigned PIRAZ/SSS duties are stationed well in advance of the main forces in the direction of the threat, often in close proximity to enemy territory. They are capable of positively identifying all friendly aircraft, serving as navigational checkpoints for strike aircraft, controlling air intercepts, coordinating search and rescue operations, maintaining plots of the air picture in CIC, and performing other duties as required to ensure protection of the force.

Phase 2: Raid Engagement

Phase 2 is a specific action against a specific threat that ends when the threat is eliminated.

COMBAT AIR PATROL (CAP)/DEFENSIVE COUNTER-AIR (DCA)
Fighter aircraft under the control of PIRAZ or other picket units are a force's first line of defense against air attack. Fighters are stationed beyond shipboard missile range and are vectored to the most favorable attack position by air intercept controllers (AIC) aboard ship. An E-2C and a group of F-14's or F/A-18's are teamed up to take advantage of the Hawkeye's powerful radar and the fighter's excellent intercept capabilities. Fighters on CAP station usually fly in fuel-conserving "racetrack" patterns. When vectored, fighters may intercept detected threats at ranges of hundreds of miles from the carrier.

AUTONOMOUS CAP/DCA
A CAP unit stationed at large distances (300 nm) from the main body, serve a dual purpose of extending the forces surveillance perimeter, usually along the threat axis, and are free to engage enemy threats based on specific Rules of Engagement (ROE).

AW Platforms and Weapons

Aircraft

Aircraft are an essential part of the AW Defense in Depth concept. Serving both as surveillance platforms and attack platforms the versatile aircraft of the U.S. fleet can meet the challenges of the intensive AW environment head on.

USN Fighter Aircraft

AIRCRAFT	F-14A TOMCAT	F/A-18A HORNET
MANUFACTURER:	Grumman	McDonnell Douglas
CREW:	Pilot, Radar-intercept Officer	Pilot
ENGINES:	2 Pratt & Whitney TF30-P-414A turbofan; 20,900 lbs ea. with afterburner	2 General Electric F404-GE-400 Turbofans;16,000 lbs ea.
WEIGHTS:	empty 58,715 lbs max. 74,349 lbs	empty 28,000 lbs; fighter takeoff 37,000 lbs; attack takeoff 48,253 lbs.
SPEED:	max. Mach 2.4	max. Mach 1.8+ at 40 Kft.
CEILING:	service 50,000+ ft.	50,000+ ft.
COMBAT RADIUS:	500 nm with combat load	415 nm—with fighter load; 550 nm—with attack load
ARMAMENT:	20-mm Vulcan cannon; combinations of Phoenix (6 max), sidewinders, & sparrow missiles	20-mm Vulcan cannon; F-sidewinder + sparrow + AMRAMM A-sidewinder + 17,000 lbs of bombs, missiles, rockets
NOTE:	An old airframe.	Has been criticized for relatively short legs.

Missiles

As with aircraft, missiles have both a name and a military designation.

Missile Designations

Launching Platform	Mission	Type
A - air	G - surface attack	M - missile
R - ship	I - aerial intercept (intercept air targets)	R - rocket
U - submarine	U - underwater attack	

For example, RIM-2E is a ship launched guided missile designed to intercept air targets. The E indicates that it is the fifth upgraded modification to the design.

Surface-to-air missiles (SAMs) are also categorized according to their range: short, medium, or long.

The most common missiles used by the U.S. Navy are the *Sparrow, Sidewinder, Phoenix, Standard Missiles (SM series), Sea Sparrow,* and *RAM.*

Sparrow (AIM-7) is a medium range semi-active radar guided weapon. It has a speed of about Mach 2.5 and carries an 88 lbs high-explosive warhead. It is an all aspect missile (i.e. it can attack an enemy aircraft from any direction) and it can operate in all types of weather. Platforms: fighter aircraft (F-14, F/A-18)

Sidewinder (AIM-9) is the most widely used air to air missile. As a heat-seeking missile, it is totally passive and requires very little equipment on the launching platform, making it a highly versatile fire and forget missile and capable of being placed on almost any aircraft. The latest models can be used against a target from any aspect. The Sidewinder has a speed of Mach 2+. It's effectiveness is reduced in bad weather. Platforms: fighter aircraft (F-14, F/A-18), attack aircraft (F/A-18, AV-8), observation aircraft (OV-10), helicopters (AH-1), others (S-3, P-3)

Phoenix (AIM-54) is only carried by the F-14 and is a "fire-and-forget" radar guided missile. Using the F-14's advanced AWG-9 radar, six of these missiles may be fired in rapid sequence at six separate targets. The missile is designed to operate in a hostile electronic countermeasures environment without having its capability degraded. The missile range is significantly longer than AIM-7 or AIM-9.

Standard missiles (SM) is a generic term designating one of three missiles all sharing a common airframe. All have semi-active radar homing guidance with a high explosive, or continuous rod warhead. All have a speed in excess of Mach 2. It has limited use in the surface to surface mode against enemy surface threats.

- SM-1 MR (RIM-66B) short range
- SM-1 ER (RIM-67A) medium range
- SM-2 MR (RIM-66C) medium range

Seasparrow (RIM-7M) The NATO Seasparrow is the Sparrow (AIM-7) missile modified for shipboard use in the basic point defense missile system (BPDMS). Although Seasparrow is designed to protect individual ships against aircraft and missile attacks, it also has limited use against surface targets. Platforms: aircraft carriers, destroyers, frigates, command ships (LCC), amphibious ships (LHA, LPH).

RAM (Rolling Airframe Missile) (RIM-116A) is a modified Sidewinder for use by surface ships as a Basic Point Defense Missile System (BPDMS) against incoming aircraft and cruise missiles. This relatively new missile is just being introduced into the fleet.

Guns

Guns are a ship's final defense against air attack. Most guns that the U.S. Navy uses today are dual-purpose (DP) guns that can be used both against air and surface targets.

The *5"/54 Caliber* automatic rapid-fire DP gun is carried by virtually all post-World War II destroyers and cruisers. The 70-pound shell has an maximum range of about 12 nm. The light weight, unmanned Mk 45 gunmount is the most common variant of this gun in the fleet today. An older manned 5" gunmount, the Mk 42, is an older version no longer used on our ships.

The *76-mm/62 Caliber Oto Melara Mk 75* rapid fire DP gun coupled with the Mk 92 GFCS is a highly accurate, lightweight gun system standard on all FFG-7, and most Coast Guard Cutters. It has a firing rate of 85 rounds per minute and a range of about 9 miles.

The *20mm Phalanx Close-In Weapon system (CIWS)* is designed to be a ship's last ditch defense against anti-ship missiles. The system includes its own search and tracking radar, a fire control system, and the 20-mm, six-barrel Vulcan cannon capable of firing 3000 rounds a minute at an effective range of up to 1 nautical mile.

Air Warfare Planning

Disposition (position of units) of a modern naval force must be flexible enough to cope with multiple, perhaps simultaneous, threats. Not only must the degree of a particular threat be taken into account, but also the force's surveillance and engagement requirements. The best formation for surveillance may not be the best for defense against air attacks. A good air defense formation may present problems in defending against enemy surface and subsurface threats. The enemy air threat may include a variety of air and surface launched missiles. Primary defense is the neutralization of launching platforms. Additionally, operating in the littoral environment presents new concerns about reaction time. Thus, flexibility is of prime importance.

Factors to consider when determining an AW disposition include:

- Coverage required (type, geographical area, threat sector/threat axis).
- Ships and aircraft available for picket duty.
- Communications capabilities and limitations.
- CAP capabilities.
- EW capabilities.
- Existing environmental conditions.

The last factor is very important, and often critical in AW. Environmental conditions can greatly affect the performance of sensors and communications, not only in the AW context, but in all other areas of naval warfare.

Positioning Forces

When forces are properly positioned, the concept of Defense in Depth can be properly employed. Within the areas of AW, various lines of defense are established in order to protect the main body or high value units. The following guidelines are utilized to position forces in accordance with the considerations above.

Stationing Aircraft

- Autonomous CAP (Fighters)
- CAP (Fighters)
- AEW (E-2C, E-3 AWACS)
- EW (EA-6B's)

Stationing Missile Ships

Medium Range Missile Platforms. Should be stationed in areas adjacent to the "Vital Area" to provide close-in protection from all directions.

Review Questions

1. Describe the concept of Defense in Depth.
2. What is the primary objective during the surveillance phase of Air Warfare?
3. What is a Picket Unit?
4. What is the mission of Combat Air Patrol in Air Warfare?
5. What is PIRAZ?
6. List three considerations that affect AW planning.

8

Strike Warfare

Learning Objectives

At the end of this lecture the student will be able to:
   Define Strike Warfare.
   State the command relationship between the CWC and the STWC.
   Describe the four mission types of Strike Warfare.
   Describe the various platforms and weapons involved is STW .
   Describe the major components of the Tomahawk Weapon System.
   Describe the advantages and disadvantages of Tomahawk and TACAIR.

Introduction

Strike Warfare is the use of tactical aircraft and/or cruise missile strikes against land targets in an offensive power projection role. During World War II, aircraft gave birth to the concept of naval strike warfare. The ability of aircraft to appear unexpectedly, anywhere within the battle space, made them particularly effective. Today, the value of naval strike warfare has increased significantly with the addition of precision-guided bombs and cruise missiles.

Strike Warfare is characterized by four missions: coordinated strike, interdiction, armed reconnaissance, and close air support. The Strike Warfare

One of eight Tomahawk cruise missiles launches from the stern of the USS *Laboon* (DDG 58) to attack selected air defense targets in Iraq on Sept. 3, 1996. The U.S. Navy Arleigh Burke class destroyer launched the missiles at 7:15 a.m. local time as it operated in the Northern Arabian Gulf. DoD photo by Petty Officer 1st Class Wayne W. Edwards, U.S. Navy.

Commander (STWC) will select one of these mission types whenever a strike is ordered. The selection of mission type is dependent upon the goals of the strike. The STWC is a subordinate commander under the Composite Warfare Commander *(CWC)* and is specifically trained in all aspects of Tactical Air (TACAIR) and Tomahawk strike capabilities.

Coordinated Strike

The coordinated strike mission (also termed deep tactical support) is to destroy specified targets at known locations. The objective of Coordinated Strike is to reduce the enemy's war-making capacity and logistic capability. Targets for this mission are likely to be well inland and highly defended. Strikes may require relatively large numbers of support aircraft to protect and assist strike aircraft. Large numbers of strike aircraft may be needed to attain reasonable levels of destruction in order to avoid having to conduct a re-strike operation.

Interdiction

The interdiction mission is to destroy specifically briefed targets which deny the enemy access to an area. It has a secondary mission of attacking targets of opportunity, if so authorized.

Armed Reconnaissance

The armed reconnaissance mission is to destroy targets of opportunity. It has a secondary mission of attacking specified fixed targets, if no target of opportunity presents itself. Armed reconnaissance is generally planned for a specified route or area, and weapons loads may be tailored to attack moving or movable targets.

Close Air Support

The close air support (CAS) mission is to harass, neutralize, or destroy enemy ground forces that present an immediate or direct threat to friendly ground forces. Close air support is provided in accordance with the supporting commander's tactical requirements and ability to control or coordinated the effort.

STW Platforms and Weapons

Attack Aircraft

The Navy's premier strike fighter for the future is the **F/A-18 E/F SUPER HORNET**. It incorporates a number of improvements over today's F/A-18 C/D including:

- 25% larger wing
- new engines
- up to 40% more range
- updated cockpit
- additional weapons stations
- reduced radar signature
- increased survivability

- growth potential

The F/A-18 E/F will be able to employ a wide variety of precision-guided weapons including laser-guided bombs and the new family of joint attack weapons (JDAM and JSOW). This important strike capability is achieved without compromising air-to-air performance.

The **F-14 TOMCAT** entered service in the mid-1970's and remains one of the premier air-to-air aircraft today. It is capable of dropping Laser-Guided Bombs (LGBs) and has proven itself as a very capable air-to-ground platform.

Air-launched Weapons

Precision-guided weapons like the Joint Stand-Off Weapon (JSOW), Joint Directed Attack Munition (JDAM), Stand-Off Land Attack Missile—Expanded Response (SLAM-ER), laser-guided bombs, and the TOMA-HAWK cruise missile make up the lethal punch of Naval Strike Warfare.

The **JOINT STAND-OFF WEAPON (JSOW)** is a joint development effort by the Navy and Air Force to produce the next generation of stand-off missiles. The Navy is the lead service in developing the weapon. The JSOW is a subsonic weapon with an approximate range of 40 nm. This missile can be armed with a 500-1000 lb warhead, BLU-108 bomblets, or Brilliant Anti-Tank (BAT) submunition. The JSOW is a GPS guided glide bomb.

The **JOINT DIRECTED ATTACK MUNITION (JDAM)** is a joint Navy-Air Force program to develop an air-launched attack munition basef on employing Mk 83 1000 lb and Mk 84 2000 lb bombs with GPS guidance kits. The JDAM is a subsonic missile with an approximate range of 12 nm. It can be used in all types of weather and does not require the aircrew to visually acquire the target prior to weapons release. Weapon targeting is performed by imagery analysts using satellite pictures of the target area.

The **STAND-OFF LAND ATTACK MISSILE-EXPANDED RE-SPONSE (SLAM-ER)** uses a Harpoon's propulsion section and warhead. The SLAM-ER is a subsonic missile with an approximate range of 120 nm.

U.S. Navy and Marine Corps Aircraft carry Mk 80 series bombs. The Mk 82 (500 lb), Mk 83 (1000 lb), and Mk 84 (2000 lb) bombs can be outfitted with laser-guided bomb kits enabling launching aircraft or other forces to guide a bomb to a target illuminated by a laser designator. Laser-guided bombs require aircrews to locate and designate the target throughout weapon flight.

Cruise Missiles

TLAM-C

(Tomahawk Land Attack Missile—Conventional) is a 1000 lb warhead installed. TLAM-C has a range of about 675 nm. Guidance for TLAM C and D are the same, both use inertial navigation and TERCOM to contour match the surface land below. Additionally, DSMAC is used to compare scenes optically viewed to an on board digital map. Block III missiles also have GPS capability. (For an explanation of TLAM terms, see below.)

TLAM-D

(Tomahawk Land Attack Missile—Bomblet Dispenser) The bomblets are numerous small shaped charges used to cover an area or several different targets. TLAM-D has a range of about 475 nm.

TLAM-N

(Tomahawk Land Attack Missile—Nuclear) is a nuclear warhead (W80 with 200-kT) attached to the Tomahawk airframe. TLAM-N has an approximate range of 1300 nm.

Tomahawk Components

DTD:

(Data Transport Device). A storage device of mission information that is loaded into the Tomahawk missile upon mission selection prior to launch. After launch the mission can not be updated.

GPS:

(Global Positioning Information). A satellite-based positioning system. The same positioning system used in electronic navigation is used in conjuction with TERCOM and/or DSMAC in the guidance of a Tomahawk cruise missile. The GPS encryption for the 'P' code is loaded into the Tomahawk during missile alignment.

APS:

(Afloat Planning System). A deployable detachment of Tomahawk mission planners and equipment that can develop, revise, and disseminate Tomahawk missions within the CV Battle Group's area of operation.

TERCOM:

(Terrain Contour Matching). A method to navigate the weapon by comparing the Tomahawks inertial position and altitude with a contour map stored in the missile's onboard computer.

DSMAC:

(Digital Scene Matching Area Correlation). Select digital scenic photographs leading to the target stored in the mission software. The missile uses an on board camera to compare photos with scenes stored in the software. This is where precise missile navigation occurs.

WARHEAD:

1000 lbs warhead, bomblet package, or nuclear warhead.

MDU:

(Mission Data Update). Missions transferred to ships via satellites and loaded into existing DTD's. Provides short term mission planning flexibility.

TAINS:

(TERCOM Aided Inertial Navigation). Basically a dead reckoning system with TERCOM updates.

The figure to the right shows a Tomahawk missile receiving preplanned mission information from shore planning facilities via the launch platform.

When the missile arrives at the first preplanned waypoint, it checks its navigation position through a series of TERCOM checks.

As the Tomahawk approaches the target, very accurate navigation is achieved through a series of DSMAC maps.

Planning Considerations

The Strike Warfare Commander must comprehend the advantages and disadvantage of using Tomahawk vs TACAIR and consider other aspects when preparing a "Strike Package."

Advantages and Disadvantages of TACAIR vs. Tomahawk

PLATFORM	ADVANTAGES	DISADVANTAGES
TACAIR	Greater payload. Target selection capability. Battle Damage Assessment (BDA). Close air support capability. Reusable.	Pilot vulnerability. Shorter legs unless refueled. Limited deep strike capability.
TOMAHAWK	Deep strike capability. No loss of pilots. Accuracy. More launch platforms spread air defenses.	No BDA No target selection capability. Small payload. High cost per shot.

Platforms and Planning Considerations

TACAIR	TOMAHAWK - TLAM	SUPPORT AIRCRAFT
How many are operationally ready?	How many platforms carry Tomahawk?	Fighters available? Air defense roll back (initial strike to reduce the SAM and AAW capability) Air defense suppression.
Are there defense requirements for SUCAP?	Do they have additional missions which would make them unavailable?	EW aircraft for chaff and jamming. AEW aircraft for tracking of return to force A/C.
How many squadrons are trained for special circumstances, ie. weather, day or night?	Are planned missions available?	SAR, RESCAP.
Distance to the target (fuel)		Reconnaissance A/C for BDA. Reconnaissance A/C escorts. Tankers.

Targeting and Weaponeering

Intelligence and Strike Warfare are inextricably linked. Intel provides crucial data for the targeting and weaponeering processes.

Targeting—The selection of targets to be struck as part of the campaign or strategy. Closely related to concepts of "critical vulnerability" and "center of gravity."

Examples: Targeting of radio relay stations to degrade enemy's C2 capability

Targeting a party headquarters to damage a country's leadership

Targeting an airfield to degrade air-defense capability.

Weaponeering—The pairing of weapons with targets to ensure maximum effectiveness and minimum risk.

Examples: Using TLAM against SAM sites that present a high risk to manned aircraft.

Using F/A-18s with LGBs against mobile units that are difficult for TLAMs to hit.

Time-Critical Targets are a growing problem. Intel and Strike Planners must work closely together to provide data for strike weapons (A/C or TLAM) in order to hit mobile targets such as SAM sites that redeploy daily in Southern Iraq.

Campaigns are almost always joint conducted under the command of the theater CINC. He designates a JFACC, who oversees all air operations, including strike operations. As part of that, his staff conducts a target prioritization process, which works as follows:

The ATO Process

Component commanders (JFMCC, JFLCC, JFACC, etc.) submit a target "wish list" to the GAT (Guidance, Apportionment, Targeting) Cell. The GAT cell merges those lists and creates a Joint Prioritized Integrated Target List (JPITL).

The component commanders also submit a list of available strike assets. Those assets are matched with the JPITL to create the Air Tasking Order (ATO).

TLAM targeting is also conducted at the theater CINC level. Once a target is designated for TLAM, a Cruise Missile Support Activity or an Afloat Planning System detachment plans the mission. Once complete the mission is transmitted to a TLAM shooter for execution.

Review Questions

1. What is the Definition of Strike Warfare?
2. What are the 4 types of Strike missions?
3. What is primary Tactical Aircraft used in Strike Missions?
4. What are the different versions of Tomahawk?

5. What is an MDU?

6. Generally, how does a Tomahawk missile find its way to the target?

7. What are the advantages and disadvantages of using strike aircraft versus Tomahawk?

8. What are some of the planning considerations when TACAIR is selected?

Suggested Further Reading

Friedman, Norman. *World Naval Weapon Systems.* Annapolis: Naval Institute Press, 1989.

Giauque, Michael, LT. "Cruising Ahead with Tomahawk." *Surface Warfare Magazine.* (September/October 1992): 8–11.

Polmar, Norman. *The Ships and Aircraft of the U.S. Fleet,* (14th ed). Annapolis: Naval Institute Press, 1987.

Prezelin, Bernard and Baker A.D. *Combat Fleets of the World 1990/1991, Their Ships, Aircraft, and Armament.* Annapolis: Naval Institute Press, 1990.

Worldwide Challenges to Strike Warfare, Office of Naval Intelligence, January 1996.

Worldwide Challenges to Strike Warfare, Office of Naval Intelligence, February 1997.

9

Expeditionary Warfare

Learning Objectives

At the end of this lecture the student will be able to:

Describe how Expeditionary Warfare doctrines contribute to the basic sea control and power projection missions of the naval service. More specifically:

Define Expeditionary Warfare.

Describe the range of military operations.

Describe the five elements of Expeditionary Warfare.

Describe the four reasons for conducting an amphibious operation.

Describe the four types of Amphibious Operations.

Describe the elements of an ATF and the functions of a LF.

Discuss the five phases of the amphibious assault in their normal sequence.

Describe the basic structure of the Amphibious Objective Area (AOA).

Additional Required Reading

None.

Introduction

The Expeditionary Warfare tradition in the United States Navy is as old as the nation itself. On March 3, 1776, Captain Samuel Nicholas landed his Continental Marine Battalion on New Providence Island in the Bahamas and seized the British fort. After independence, U.S. Navy forces successfully eliminated Barbary Coast extortion in the early 1800's, establishing the precedent for global naval expeditionary operations.

Modern naval Expeditionary Warfare had its origins in exercises conducted during the decades preceding World War II. The exercises identified unique operational and equipment requirements and developed the skills that were perfected in carrier and amphibious operations, which ensured victory in the Pacific Theater.

Modern Expeditionary Warfare

Today, Expeditionary Strike Groups (ESG) carries on the expeditionary tradition of "Maneuver Warfare." They provide the nation a fully ready combat service for forward deployed and able to respond at a moments notice. Naval Expeditionary Forces were used to restore democracy in Haiti in 1993–1994 and to protect U.S. citizens and diplomatic personnel in Monrovia, Liberia, in April 1996. Strikes from the sea punished Iraq for violating United Nation resolutions in 1993 and 1996.

Ready for combat, Naval Expeditionary Forces are simultaneously capable of performing large-scale humanitarian operations without reconfiguration. En route home from Operation Desert Storm, Amphibious Ready Group Three was diverted to Bangladesh in 1991, which had been devastated by a tropical cyclone. Navy and Marine Corps Expeditionary Forces protected U.S. citizens in Somalia in 1992 and covered U.N. forces in 1993 and 1995. In late 2001, Marine units deployed from amphibious forces have fought the Taliban and Al-Qaeda in Afghanistan, demonstrating the reach of the Navy and Marine Corps team into landlocked regions as well as coastal ones. Most recently, two Amphibious Task Forces (ATF), embarking two Marine Expeditionary Brigades (MEB), were stationed off the coast of Iraq as part of Operation Iraqi Freedom. The flexibility these forces allowed the Marines to swiftly stage personnel and equipment in Kuwait and surge into Iraq once the war commenced. All the while, the Marine Air Wings were able to maintain 24/7 operations from the large deck amphibious ships offshore (Sea Basing).

Lance Corporal R.A. Black from L3/2 Landing Zone Security Guard for the AV-8A Harrier Aircraft of VMA-513, from Beaufort S.C. looking on as a Harrier is launched on a bombing mission from Landing Zone Goose, Camp Lejeune, N.C. during exercise Versatile Warrier in which the Harrier was tested extensively.

Naval Expeditionary Warfare comprises military operations mounted from the sea, usually on short notice consisting of forward deployed, or rapidly deployable, self-sustaining naval forces tailored to achieve a clearly stated objective. Additionally, Expeditionary Warfare Forces support and perform the full range of military operations on a day-to-day basis.

Naval Expeditionary Forces retain the unique ability to quickly shift operational focus across the spectrum—from combat missions to humanitarian operations—without needing to reconfigure the force.

RANGE OF MILITARY OPERATIONS

Peacetime Engagement	Deterrence and Conflict Prevention	Fight and Win
*Military-to-Military Contact	*Regional Alliances	*Clear Objectives—Decisive Force
*Assistance to Nations	*Crisis Response	*Wartime Power Projection
*Humanitarian Operations	*Confidence-Building Measures	*Combined and Joint Warfare
*Peacekeeping	*Non-combatant Evacuation Operations	*Winning the Information War
	*Sanctions Enforcement	*Counter weapons of mass destruction
	*Peacekeeping	*Force Generation
		*Winning the Peace

There are five principal elements that are vital to mission success in Expeditionary Warfare:

Maritime Dominance—To approach land in order to execute an Expeditionary Mission, U.S. Navy and Marine Corps forces must operate in shallow and restricted waters. Control of the sea en route and control of the littoral waters in a large area surrounding the objective is vital to mission success. To get to the objective, our forces will have increasingly fewer forward bases from which to stage. Developments in surveillance will make it increasingly difficult to transit and remain undetected and untargeted. Shallow and restricted waters provide and arena for submarines, torpedoes, and next generation anti ship cruise missiles. Mines (responsible for the loss or damage of more U.S. Navy ships in recent years than all other weapons combined) will be found throughout the range of coastal waters. Elements of Sea Power 21 are specifically designed to alleviate the threat to our forces as the approach and operate in the littoral regions.

Firepower—To be able to detect, identify, categorize, and destroy a target at will is the goal of firepower. It requires sophisticated capabilities in surveillance, fire delivery, and maneuver. Artillery is more self-contained and mobile, enabling artillery and theater ballistic missile batteries to change positions in a few hours, with setup times in minutes. Land based artillery and rockets often outrange naval guns and can deliver conventional explosives, mines or precision guided munitions. In the next 20 years, laser and energy technologies will transform today's sensor-blinder into a hard-kill weapon. Theater Ballistic Missiles (TBM), already a threat to Naval Expeditionary Forces at the beachhead or in ports, will become accurate enough to target individual ships in amphibious landing areas. Concentration of hostile firepower against Expeditionary Forces will make the environment more challenging. Ongoing developments of ships capable of countering the TBMs are under have been tested and should be operational in the near future.

Maneuver Dominance—To gain the initiative critical to mission success, the Expeditionary Commander must have the freedom to maneuver at sea and on land. Natural and man-made obstacles, poor infrastructure, harsh climates, and the combat capabilities of highly mobile armored reaction forces can restrict our ability to maneuver. A turbulent human environment across the range of military operations further limits options. The Marine Corp and Navy are striving to develop ships and systems (Transformation Initiatives) that will afford more flexibility in a constrained environment.

Air Dominance—Control of the air over an objective is mandatory, yet potentially unstable nations are importing or building increasingly sophisticated SAM's and integrating them into modern air defense networks. They are also acquiring advance aircraft necessary to patrol air space.

Information Superiority—The commander remains dependent upon the quality and quantity of information that the Command, Control, Communications, Computers, Intelligence, Surveillance, and Reconnaissance assets can deliver. Potential aggressors are using stealth technologies and camouflage, concealment, and deception techniques to hide potential targets. Opposing naval forces blend into the background and clutter; key military facilities and capabilities are buried both physically in the earth, and figuratively within urban areas. Our systems seek them out better than ever before in history, but at the price of demanding a geometric increase in data transfer rates. Communications rely upon satellite links, more efficient use of the electromagnetic spectrum, sophisticated encryption techniques, new transmission devices, and other innovative technologies. All are vulnerable.

One of the principal means of conducting Expeditionary Warfare is through the Amphibious Operation; and in order to be successful, the Amphibious Operation must employ each of the five elements of Expeditionary Warfare.

Amphibious Operations

An amphibious operation is a military operation launched from the sea by naval and landing forces embarked in ships or craft involving a landing on a hostile or potentially hostile shore. It is directed by the combatant commander, subunified commander, or JTF commander delegated overall responsibility for the operation. An amphibious operation requires extensive air participation and is characterized by closely integrated efforts of forces trained, organized, and equipped for different combat functions.

Amphibious operations are designed and conducted primarily to:

- *Prosecute further combat operations.*
- *Obtain a site for an advanced naval, land, or air base.*
- *Deny use of an area or facilities to the enemy.*
- *Fix enemy forces and attention, providing opportunities for other combat operations.*

The essential usefulness of an amphibious operation stems from its mobility and flexibility (i.e., the ability to concentrate balanced forces and strike with great strength at a selected point in the hostile defense system). The amphibious operation exploits the element of surprise and capitalizes on enemy weaknesses by projecting and applying combat power at the most advantageous location and time. The threat of an amphibious landing can induce enemies to divert forces, fix defensive positions, divert major resources to coastal defenses (Operation Desert Storm), or disperse forces. Such a threat may result in attempting to defend their coastlines.

The salient requirement of an amphibious assault is the necessity for swift, uninterrupted buildup of sufficient combat power ashore from an initial zero capability to full, coordinated striking power as the attack progresses toward the amphibious task force's (ATF) final objectives.

Types of Amphibious Operations

AMPHIBIOUS ASSAULT—This is the principal type of amphibious operation, which is distinguished from other types of amphibious operations in that it involves establishing a force on a hostile or potentially hostile shore (Inchon, Korea).

AMPHIBIOUS WITHDRAWAL—An amphibious operation involving the extraction of forces by sea in naval ships or craft from a hostile or potentially hostile shore.

AMPHIBIOUS DEMONSTRATION—An amphibious operation conducted to deceive the enemy by a show of force with the expectation of deluding the enemy into a course of action unfavorable to it.

AMPHIBIOUS RAID—An amphibious operation involving swift incursion into or temporary occupation of an objective followed by a planned withdrawal. Raids are conducted for such purposes as:

- Inflicting loss or damage.
- Securing information.
- Creating a diversion.
- Capturing or evacuating individuals and/or material.
- Destroying enemy information gathering systems to support operations security.

The Initiating Directive

The *initiating directive* is the establishing order to the ESG to conduct an amphibious operation. It is used by the combat commander, subunified commander, service component commander, or Joint Task Force (JTF) commander to establish and assign mission parameters such as mission assignments, area of operation, codes, target dates, and special instructions. It may not be a comprehensive document.

Chain of Command

The interrelationship of the Navy and the embarked Marines (MEU Staff) during the planning and execution of an amphibious operation requires the establishment of parallel chains of command and corresponding commanders at all levels of the Expeditionary Strike Group (ESG) organization. Fundamental considerations governing the application of such a system of parallel command include:

- *Commander, Expeditionary Strike Group,* is overall responsible for the planning and execution of all naval ship movement from the staging area to the Amphibious Objective Area (AOA). In addition, the ESG Commander directly oversees all ship to shore movements of personnel and equipment.

    Currently, the command structure of the ESG is under study. The West Coast is experimenting with a Flag Officer as being designated as the ESG Commander and is alternating that post between a Navy and Marine Corps Officer. If this model proves successful, it will be adopted on the East Coast.
- *Commanding Officer, Marine Expeditionary Unit (MEU),* is normally a 0-6 Marine Corps officer, who has Operational Control (OPCON) of the Landing Force (LF). With a traditional ESG (East Coast), the MEU Commander has a co-equal relationship with the ESG Commander on all matters pertaining to the planning and execution of any operations involving the LF. With the West Coast model, the MEU Commander works closely with but is subordinate to the Flag Officer.

The Navy and Marine Corps enjoy a close relationship for planning purposes, and must work together to ensure a good operation. Each brings his/her expertise to the most complex military operation.

Today's Navy/Marine Corps working relationship within an ESG is typified by what is called the "Supported-Supporting" relationship. Simply stated, the Supported Commander is the one who has been designated overall in command of any particular mission. The Supporting Commander will in turn offer all reasonable assistance in order to ensure operational success.

Examples:

1. If the MEU Commander is tasked to conduct a NEO (Non Combatant Evacuation Operation), the ESG Commander will render all necessary assistance to the MEU. This would obviously include the stationing of the amphibious ships, ship to shore movement, etc.
2. If the ESG Commander is tasked with conducting a MIO (Maritime Interdiction Operation), the MEU Commander will in turn make assets available if so requested. Examples of this might include Marine personnel and or helicopters.

Note: With the current Flag Officer configuration with the West Coast ESGs, there are some significant modifications in the Command and Control relationships. Fundamentally, however, the Supported-Supporting concept is still in play.

Amphibious Task Force Organization

The organization for execution of the amphibious operation reflects interrelationships at every level between the tasks of the LF, corresponding naval forces, special operations forces (SOF), and participating Air Force forces. The interrelationships dictate that special emphasis be given to the parallel chains of command.

The task organization of the ATF must meet the requirements of embarkation, movement to the AOA, protection, landing, and support of the LF. For this reason, the task organization is determined according to the requirements of the anticipated tactical situation. An ESG in most scenarios is sufficient for mission accomplishment; however, a larger force might be required; flexibility is essential.

Amphibious Task Force

The task organization formed for conducting an amphibious operation is the ATF. The ATF always includes Navy forces and a LF, both of which normally have organic aviation assets. Other air and Special Operations Forces (SOF) may be included, as required. Task elements of the ATF are listed below:

- *Transport Groups*
- *Control Group (ship-shore movements)*
- *TACAIR Control Group*
- *Naval Surface Fire Support (NSFS) Group*
- *Carrier Battle Group/Warfare Commanders*
- *TACAIR Group (shore based)*
- *Mine Warfare Groups*
- *SPECWARFARE Groups*
- *Tactical Deception Groups*
- *Naval Beach Group*
- *Construction Battalions*

Landing Forces

The LF consists of the command, combat, combat support, and combat service support (CSS) elements assigned to conduct the amphibious assault (air and ground). The LF is specially organized for the following functions:

- *Embarkation of troops, equipment, and supplies.*
- *Debarkation and landing of troops by air and/or surface units.*
- *Conduct of air and waterborne assault operations.*
- *Control of Naval Surface Fire Support (NSFS).*
- *Provision, as appropriate, and control of air support.*
- *Operation and tactical employment of organic amphibious vehicles and aircraft.*
- *Discharge of logistics and CSS elements and cargo from assault shipping and establishment of logistical sites and service areas.*

Phases of the Amphibious Assault

The amphibious assault follows a well defined pattern. The general sequence consists of planning, embarkation, rehearsal, movement to the landing area, assault, and accomplishment of the ATF mission. While planning occurs throughout the entire operation, it is normally dominant in the period before embarkation. Successive phases bear the title of the dominant activity taking place within the period covered.

The organization of embarkation needs to provide for maximum flexibility to support alternate plans that may of necessity be adopted. The landing plan and the scheme of maneuver ashore are based on conditions and enemy capabilities existing in the Amphibious Objective Area (AOA) before embarkation of the LF.

Planning

The *planning phase* denotes the period extending from the issuance of the initial directive to embarkation. Although planning does not cease with the termination of this phase, it is useful to distinguish the change that occurs in the relationship between commanders at the time the planning phase terminates and the operational phase begins. Logistics is the science of planning and carrying out the movement and maintenance of forces. In an amphibious operation logistics deals with the design, development, acquisition, storage, movement, distribution, maintenance, evacuation, and disposition of material and personnel. Combat Service Support (CSS) provides the essential logistics functions and tasks necessary to sustain all elements of operating forces in an area of operation. Planning considerations include assembly and embarkation of personnel and material based on anticipated requirements of LF scheme of maneuver ashore. All considerations that lead to a successful operation should be dealt with in the planning phase, but may be modified by the results of the rehearsal. These considerations include, but are not limited to: the assembly and embarkation of personnel and material based on the anticipated movement ashore (*first off, last on*), the anticipated strength of the enemy, climate and terrain of the area of operations, the anticipated length of supply lines, communications capabilities, and target dates. Effective logistics and combat service support are absolutely critical to the success of any amphibious operation.

Embarkation

The *embarkation* phase is the period during which the forces, with their equipment and supplies, embark in assigned shipping.

Rehearsal

The *rehearsal phase* is the period during which the prospective operation is rehearsed for the purpose of:

• Testing the adequacy of the plans, the timing of detailed operations, and the combat readiness of participating forces.
• Ensuring all echelons are familiar with plans.
• Testing communications.

Movement

The *movement phase* is the period during which various elements of the ATF move from points of embarkation to the AOA. This move may be via rehearsal, staging, or rendezvous areas. This movement phase is completed when the various elements of the ATF arrive at their assigned positions in the AOA.

Assault

The *assault phase* begins when sufficient elements of the main body of the ATF arrive in assigned positions in the landing area and are capable of beginning the ship-to-shore movement. The assault phase terminates with accomplishment of the ATF mission. The assault phase encompasses:

- Preparation of the landing area by supporting arms.
- Ship-to-shore movement of the LF.
- Air and surface assault landing (assault and initial unloading; tactical) by assault elements of the LF to seize the beachhead and designates ATF and LF objectives.
- Provision of supporting arms and logistics/CSS throughout the assault.
- Landing the remaining elements (general unloading; logistical) for conduct of operations as required for accomplishment of the ATF mission.

PERMA vs. EMPRA

PERMA is the normal doctrinal sequence for conducting an amphibious operation. However, when the organization involved is the ESG and the MEU(SOC), the actual sequence of events that might occur is EMPRA. Simply put, the MEU(SOC) embarks aboard ESG shipping, moves to the area of operations, then receives a mission and conducts the planning, the rehearsal, and the assault. This is due to the forward deployed nature of the ESG/MEU, which is scheduled months in advance.

Amphibious Objective Area

The Amphibious Objective Area (AOA) is a geographical area, delineated in the initiating directive, for purposes of command and control, within which is located the objectives to be secured by the amphibious task force. This area must be of sufficient size to ensure accomplishment of the amphibious task force's mission and must provide sufficient area for conducting necessary sea, air and land operations.

Movement of the ATF to the AOA includes departure of ships from loading points in an embarkation area; passage at sea; and approach to, and arrival in, assigned positions in the AOA. The ATF is organized for movement into movement groups, which sail in accordance with the movement plan on prescribed routes. Protective measures are utilized to prevent losses enroute. These most often entail combatant escorts and attention to EMCON, tactical evasion and sometimes deception procedures. Movement of the ATF to the AOA may be interrupted by rehearsals, diversion to staging areas for logistics/CSS reasons, or temporary stops at regulating stations or points.

Echelons of the Landing Force

Assault Echelon (AE)
The element of a force that is scheduled for initial assault on the objective area. The AE is embarked on amphibious assault shipping and comprises the tailored units and equipment packages along with maximum amount of supplies that can be loaded to sustain the assault.

Sea Echelon (SE)
A sea echelon is a portion of the assault shipping that withdraws from, or remains out of the transport area during an amphibious landing and operates in designated areas seaward in an on-call or unscheduled status.

Areas Within the AOA

Landing Beach
That portion of the shoreline over which a force approximately the size of a Battalion Landing Team may be landed.

Boat Lane
Lanes in which landing craft and AAVs proceed toward the beach.

NSFS Area
An area in which supporting ships provide cover using missiles and guns.

Helicopter Landing Area (HLA)
A specified ground area for landing assault helicopters to embark or disembark troops and/or cargo. It may contain one or more landing sites.

Transport Area
In amphibious operations, an area assigned to a transport organization for the purpose of debarking troops and equipment. It consists of mineswept lanes, areas, and channels leading to the beaches. The maximum number of ships in the transport area is directly limited by dispersion requirements, availability of forces for MCM operations, and local hydrography and topography.

Sea Echelon Area

An area to seaward of the transport area from which assault shipping is phased into the transport area and to which assault shipping withdraws from the transport area.

Operations in the AOA

The CATF controls the assault from the Task Force Command Center aboard the most capable amphibious ship (LCC, LHD, LHA, LPD). He exercises control through the following control centers and group commanders:

SUPPORTING ARMS COORDINATION CENTER (SACC):

- Coordinates pre-assault air strikes, beach clearance operations and gunfire
- Coordinates NSFS, artillery support during assault
- Coordinates helo assault through Helo Coordination Section (HCS)
- Coordinates close air support strikes thru Tactical Air Coordination Center (TACC) and AAW over transport area with the AAWC.
- SACC functions shift to Fire Support Control Center (FSCC) ashore under the CLF's control. CLF establishes a TACC, HCS, logistics control centers, and others before assuming full responsibility for future operations ashore.

BOAT GROUP COMMANDER

- Controls movement of landing craft to and from the beach.

TRANSPORT GROUP COMMANDER

- Controls movement and off loading of transports, LST's, and Assault Ships and assigns Fire Support Areas for NSFS ships.

ASW/AAW/ASUW GROUP COMMANDERS

- Conduct screening operations in the Sea Echelon Area

Tactical Considerations

Natural hazards

Surf, weather, and hydrography can often be overcome by careful planning and use of special tactics such as vertical assault, LCAC's, underwater demolition, careful timing, etc.

Vulnerability

The landing force is extremely vulnerable in the early hours of the assault as strength ashore must be built up from zero. Careful planning and coordinated execution of the plan is required. The buildup must be rapid and uninterrupted.

Battlespace Dominance

Control of sufficient land, sea, and air space to gain combat superiority in the AOA during the landing is critical. Enemy airfields, shore based cruise missile sites, and long range artillery must be neutralized by the pre-assault strikes and bombardment. The presence of mines and defensive cruise missiles may force the amphibious assault force to keep a longer standoff range and conduct an over-the-horizon assault.

Review Questions

1. What is Maritime Dominance and how can it be achieved?
2. List two reasons for conducting an Amphibious Operation?
3. What is a raid and how is it different from an assault?
4. What is the purpose of Initiating Directive as it relates to planning an amphibious operation?
5. Describe the command relationship between the CATF and CLF.
6. List four of the elements that make up an ATF.
7. What are the phases of the amphibious assault in their normal sequence?
8. Describe the areas/zones of the Amphibious Objective Area.
9. Why is Battlespace Dominance an important tactical consideration in Amphibious Operations?

Suggested Further Reading

Joint Pub 3-02: *Joint Doctrine for Amphibious Operations* (1992) Government Printing Office, Washington D.C.

NAVEDTRA 10776-A: *Surface Ship Operations* Naval Education and Training Command Government Printing Office, Washington D.C.

Wadle, S. Midn, USN *Operation Roman Candle* [NS 310 Project], U. S. Naval Academy Annapolis MD

Worldwide Challenges to Naval Expeditionary Warfare, Office of Naval Intelligence, March 1997.

10

Marine Corps Warfighting

Learning Objectives

At the end of this lecture the student will be able to:
    Explain the following:
      Maneuver Warfare
      Decision Making Process
      Philosophy of Command
      Mission Orders
      Commander's Intent
    Describe the MAGTF concept and the different types of MAGTF.
    Demonstrate how to estimate a situation using METT-T
    Utilize the concepts of maneuver warfare in a Tactical Decision Game (TDG).

Additional Required Reading

None.

Introduction

To understand the Marine Corps' philosophy of warfighting, we first need an appreciation for the nature of war itself. War is a state of hostilities that exists between or among nations, characterized by the use of military force. The aim in war is to impose our will on the enemy. We must either eliminate his physical ability to resist, or short of this, we must deploy his will to resist. The sole justification for the United States Marine Corps is to secure or protect national policy objectives by military force when peaceful means alone cannot.

Warfighting

The challenge is to identify and adopt a concept of warfighting consistent with out understanding of the nature and theory of war and the realities of the modern battlefield. This requires a concept of warfighting that will function in an uncertain, chaotic, and fluid environment—in fact, one that will exploit these conditions to advantage. It requires a concept that generates and exploits superior tempo and velocity. It requires a concept that is consistently effective across the full spectrum of conflict, because we can not attempt to change our basic doctrine from situation to situation and expect to be proficient. It requires a concept which recognizes and exploits the fleeting opportunities which naturally occur in war. It requires a concept which takes in to account the moral as well as the physical forces of war. It requires a concept with which we can succeed against a numerically superior foe, because we can no longer presume a numerical advantage. And, especially in expeditionary situations in which public support for military action may be tepid and short lived, it requires a concept with which we can win quickly against a larger foe on his home soil, with minimal casualties and external support.

Maneuver Warfare

The Marine Corps concept for winning under these conditions is a warfighting doctrine based on rapid, flexible, and opportunistic maneuver. But in order to fully appreciate what we mean by *maneuver* we need to clarify the term. The traditional understanding of maneuver is a spatial one; that is, we maneuver in space to gain a positional advantage. However, in order to maximize the usefulness of maneuver, we must consider maneuver *in time* as well; that is, we generate a faster operational tempo than the enemy to gain a temporal advantage. It is through maneuver in

South Korea . . . A Marine wearing an M-17A1 field protective mask stands ready with his M-249 squad automatic weapon during the combined South Korean/U.S. exercise Team Spirit '90.

dimensions that an inferior force can achieve decisive superiority at the necessary time and place.

From this definition we see that the aim in maneuver warfare is to render the enemy incapable of resisting by shattering his moral and physical cohesion-his ability to fight as an effective, coordinated whole—rather than to destroy him physically through incremental attrition, which is generally more costly and time-consuming. Ideally, the components of his physical strength that remain are irrelevant because we have paralyzed his ability to use them effectively. Even if an outmaneuvered enemy continues to fight as individuals or small units, we can destroy the remnants with relative ease because we have eliminated his ability to fight effectively as a force.

This is not to imply that firepower is unimportant. On the contrary, the suppressive effects of firepower are essential to our ability to maneuver. Nor do we mean to imply that we will pass up the opportunity to physically destroy the enemy. We will concentrate fires and forces at decisive points to destroy enemy elements when the opportunity presents itself and when it fits our larger purposes. But the aim is not an unfocused application of firepower for the purpose of incrementally reducing the enemy's physical strength. Rather, it is the *selective* application of firepower in support of maneuver to contribute to the enemy's shock and moral disruption. The greatest value of firepower is not physical destruction—the cumulative effects of which are felt only slowly— but the moral dislocation it causes.

If the aim of maneuver warfare is to shatter the enemy's cohesion, the immediate object toward that end is to create a situation in which he cannot function. By our actions, we seek to pose menacing dilemmas in which events happen unexpectedly and faster than the enemy can keep up with them. The enemy must be made to see his situation not only as deteriorating, but deteriorating at an ever-increasing rate. The ultimate goal is panic and paralysis, an enemy who has lost the ability to resist.

Inherent in maneuver warfare is the need for speed to seize the initiative, dictate the terms of combat, and keep the enemy off balance, thereby increasing his friction. Through the use of greater tempo and velocity, we seek to establish a pace that the enemy cannot maintain so that with each action his reactions are increasingly late—until eventually he is overcome by events.

Also inherent is the need for violence, not so much as a source of physical attrition but as a source of moral dislocation. Toward this end, we concentrate strength against *critical* enemy vulnerabilities, striking quickly and boldly where, when, and how it will cause the greatest damage to our enemy's ability to fight. Once gained or found, any advantage must be pressed relentlessly and unhesitatingly. We must be ruthlessly opportunistic, actively seeking out signs of weakness, against which we will direct all available combat power. And when the *decisive* opportunity arrives, we must exploit it fully and aggressively, committing every ounce of combat power we can muster and pushing ourselves to the limits of exhaustion.

The final weapon in our arsenal is surprise, the combat value of which we have already recognized. By studying our enemy we will attempt to appreciate his perceptions. Through deception we will try to shape his expectations. Then we will dislocate them by striking at an unexpected time and place. In order to appear unpredictable, we must avoid set rules and patterns, which inhibit imagination and initiative. In order to appear ambiguous and threatening, we should operate on axes that offer several courses of action, keeping the enemy unclear as to which we will choose.

Philosophy of Command

It is essential that our philosophy of command support the way we fight. First and foremost, *in order to generate the tempo of operations we desire and to best cope with the uncertainty, disorder, and fluidity of combat, command must be decentralized.* That is, subordinate commanders must make decisions on their own initiative, based on their understanding of their senior's intent, rather than passing information up the chain of command and waiting for the decision to be passed down. Further, a competent subordinate commander who is at the point of decision will naturally have a better appreciation for the true situation that a senior some distance removed. Individual initiative and responsibility are of paramount importance. The principal means by which we implement decentralized control is through the use of mission tactics, which we will discuss in detail later.

Second, since we have concluded that war is a human enterprise and no amount of technology can reduce the human dimension, our philosophy of command must be based on human characteristics rather than on equipment or procedures. Communications equipment and command staff procedures can enhance our ability to command, but they must not be used to replace the human element of command. Our philosophy must not only accommodate but must exploit human traits such as boldness, initiative, personality, strength of will, and imagination.

Our philosophy of command must also exploit the human ability to communicate *implicitly*. We believe that *implicit communication*—to communicate through *mutual understanding,* using a minimum of key, well-understood phrases or even anticipating each other's thoughts—is a faster, more effective way to communicate than through the use of detailed, explicit instructions. We develop this ability through familiarity and trust, which are based on a shared philosophy and shared experience.

This concept has several practical implications. First, we should establish long-term working relationships to develop the necessary familiarity and trust. Second, key people—"actuals"—should talk directly to one another when possible, rather than through communicators or messengers. Third, we should communicate orally when possible, because we communicate also in *how* we talk; our inflections and tone of voice. And fourth, we should communicate in person whenever possible, because we communicate also through our gestures and bearing.

A commander should command from well forward. This allows him to see and sense firsthand the ebb and flow of combat, to gain an intuitive appreciation for the situation which he cannot obtain from reports. It allows him to exert his personal influence at decisive points during the action. It also allows him to locate himself closer to the events that will influence the situation sot that he can observe tem directly and circumvent the delays and inaccuracies that result form passing information up the chain of command. Finally, we recognize the importance of personal leadership. Only by his physical presence—by demonstrating the willingness to share danger and privation—can the commander fully gain the trust and confidence of his subordinates.

As part of our philosophy of command we must recognize that war is inherently disorderly, uncertain, dynamic, and dominated by friction. Moreover, maneuver warfare, with its emphasis on speed and initiative, is by nature a particularly disorderly style of war. The conditions ripe for exploitation are normally also very disorderly. For commanders to try to gain certainty as a basis for actions, maintain positive control of events at all times, or shape events to fit their plans is to deny the very nature of war. We must therefore be prepared to cope—even better, to *thrive*—in an environment of chaos, uncertainty, constant change, and friction. If we can come to terms with those conditions and thereby limit their debilitating effects, we can use them as a weapon against a foe who does not cope as well.

In practical terms this means that we must not strive for certainty before we act for in so doing we will surrender the initiative and pass up opportunities. We must not try to maintain positive control over subordinates since this will necessarily slow our tempo and inhibit initiative. We must not attempt to impose precise order to the events of combat since this leads to a formulistic approach to war. And we must be prepared to adapt to changing circumstances and exploit opportunities as they arise, rather than adhering insistently to predetermined plans.

There are several points worth remembering about our command philosophy. First, while it is based on our warfighting style, this does not mean it applies only during war. We must put it into practice during the preparation for war as well. We cannot rightly expect our subordinates to exercise boldness and initiative in the field when they are accustomed to being over supervised in the rear. Whether the mission is training, procuring equipment, administration, or police call, this philosophy should apply.

Next, our philosophy requires competent leadership at all levels. A centralized system theoretically needs only one competent person, the senior commander, since his is the sole authority. But a decentralized system requires leaders at all levels to demonstrate sound and timely judgment. As a result, initiative becomes an essential condition of competence among commanders.

Our philosophy also requires familiarity among comrades because only through a shared understanding can we develop the implicit communication necessary for unity of effort. And, perhaps most important, our philosophy demands confidence among seniors and subordinates.

Decision-making

Decision-making is essential to the conduct of war since all actions are the result of decisions- or of nondecisions. If we fail to make a decision out of lack of will, we have willingly surrendered the initiative to our foe. If we consciously postpone taking action for some reason, that is a decision. Thus, as a basis for action, any decision is generally better than no decision.

Since war is a conflict between opposing wills, we cannot make decisions in a vacuum. We must make our decisions in light of the enemy's anticipated reactions and counteractions, recognizing that while we are trying to impose our will on our enemy, he is trying to do the same to us.

Whoever can make and implement his decisions consistently faster gains a tremendous, often decisive advantage. Decision making thus becomes a time-competitive process, and timeliness of decisions becomes essential to generating tempo. Timely decisions demand rapid thinking, with consideration limited to essential factors. We should spare no effort to accelerate our decision-making ability.

A military decision is not merely a mathematical computation. Decision making requires both the intuitive skill to recognize and analyze the essence of a given problem and the creative ability to devise a practical solution. This ability is the product of experience, education, intelligence, boldness, perception, and character.

We should base our decisions on *awareness* rather than on mechanical *habit*. That is, we act on a keen appreciation for the essential factors that make each situation unique instead of form conditioned response.

We must have the moral courage to make tough decisions in the face of uncertainty- and accept full responsibility for those decisions—when the natural inclination would be to postpone the decision pending more complete information. To delay action in an emergency because of incomplete information shows a lack of moral courage. We do not want to make rash decisions, but we must not squander opportunities while tying to gain more information.

We must have the moral courage to make bold decisions and accept the necessary degree of risk when the natural inclination is to choose a less ambitious tack, for "in audacity and obstinacy will be found safety."

Finally, since all decisions must be made in the face of uncertainty and since every situation is unique, there is no perfect solution to any battlefield problem. Therefore, we should not agonize over one. The essence of the problem is to select a promising course of action with an acceptable degree of risk, and to do it more quickly than our foe. In this respect, "a good plan violently executed *now* is better than a perfect plan executed next week."

Mission Tactics

Having described the object and means of maneuver warfare and its philosophy of command, we will next discuss how we put maneuver warfare into practice. First is through the use of mission tactics. Mission tactics are just as the name implies: the tactic of assigning a subordinate mission without specifying how the mission must be accomplished. We leave the manner of accomplishing the mission to the subordinate, thereby allowing him the freedom—and establishing the duty—to take whatever steps he deems necessary base on the situation. The senior prescribes the method of execution only to the degree that is essential for coordination. It is this freedom for initiative that permits the high tempo of operations that we desire. Uninhibited by restrictions from above, the subordinate can adapt his actions to the changing situation. He informs his commander what he has done, but he does not wait for permission.

It is obvious that we cannot allow decentralized initiative without some means of providing unity, or focus, to the various efforts. To do so would be to dissipate our strength. We seek unity, not through imposed control, but through *harmonious* initiative and lateral coordination.

Commander's Intent

We achieve this harmonious initiative in large part through the use of the commander's intent. There are two parts to a mission: the task to be accomplished and the reason, or intent. The task describes the desired result of the action. Of the two, the intent is predominant. While a situation may change, making the task obsolete, the intent is more permanent and continues to guide our actions. Understanding our commander's intent allows us to exercise initiative in harmony with the commander's desires.

In order to maintain our focus on the enemy, we should try to express intent in terms of the enemy. The intent should answer the question: *What do I want to do to the enemy?* This may not be possible in all cases, but it is true in the vast majority. The intent should convey the commander's *vision.* It is not satisfactory for the intent to be "to defeat the enemy." To win is always our ultimate goal, so an intent like this conveys nothing.

From this discussion, it is obvious that a clear explanation and understanding of intent is absolutely essential to unity of effort. It should be a part of any mission. The burden of understanding falls on senior and subordinate alike. The senior must make perfectly clear the result he expects, but in such a way that does not inhibit initiative. Subordinates must have a clear understanding of what their commander is thinking. Further, they should understand the intent of the commander two levels up. In other words, a platoon commander should know the intent of his battalion commander, or a battalion commander the intent of his division commander.

Marine Corps Organization—The MAGTF Concept

Marine Air-Ground Task Force (MAGTF)

The MAGTF is the force that the Marine Corps employs to conduct maneuver warfare in a combined arms operation. The nature of the MAGTF—cohesion, unity of effort, flexibility, and self-sustainment—

makes it equal to the requirements of combined arms warfare. The MAGTF contains four elements that can be tailored to a combined arms operation: a command element, a ground combat element, an aviation combat element, and a combat service support element. The MAGTF draws forces from ground, aviation, and combat service support organizations of the Fleet Marine Force (FMF) to meet this requirement.

Six Special Core Competencies

MAGTF operations are built on a foundation of six special core competencies:

a) Expeditionary readiness
b) Expeditionary operations
c) Combined-arms
d) Forcible entry from the sea
e) Sea based operations
f) Reserve integration:
    (1) Other units (MAGTFs)
    (2) Joint or coalition forces

Types of MAGTFs

MAGTFs range in size from the smallest (which can number from fewer than 100 to 3,000 Marines) to the largest (which can number from 40,000 to 100,000 Marines). There are four basic sizes/types of MAGTFs

a) Marine Expeditionary Force (MEF)
b) Marine Expeditionary Brigade (MEB)
c) Marine Expeditionary Unit (MEU)
d) Special Purpose MAGTF (SPMAGTF)

MAGTF Composition

Regardless of the size of the MAGTF, all have the same basic structure. There are four elements of a MAGTF: the command element (CE), the ground combat element, (GCE), the aviation combat element (ACE), and the combat service support element (CSSE).

a) Command Element (CE). The command element is task organized to provide command and control capabilities (including intelli-

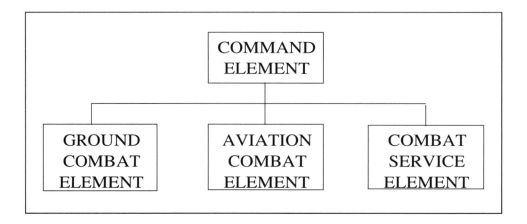

gence and communications) necessary for effective planning, direction, and execution of all operations.

(1)   Composition of the CE

    (a) MAGTF Commander

    (b) Deputy Commander

    (c) General Staff

    (d) Special Staff

(2)   Functions of the CE. Several key aspects of the CE activities are different from those of its major subordinate commands.

    (a) Drive operations

    (b) Requesting and integrating joint capabilities

    (c) Collecting intelligence

    (d) Deep, close, and rear operations

    (e) Deception and psychological operations

    (f) NBC weapon systems

    (g) Command, control, communications and intelligence

    (h) MAGTF concept of operations

    (i) Task organizing the MAGTF forces

b)   Ground Combat Element (GCE). The GCE is task-organized to conduct ground operations in support of the MAGTF mission. It is normally formed around an infantry organization reinforced with requisite artillery, reconnaissance, armor, and engineer forces and can vary in size and composition from a rifle platoon to one or more Marine divisions. It has some organic combat service support capability.

c)   Aviation Combat Element (ACE). The ACE is task-organized to support the MAGTF mission by performing some or all of the six functions of Marine aviation. It is normally built around an aviation organization that is augmented with appropriate air command and control, combat, combat support, and CSS units. The ACE can operate effectively from ships, expeditionary airfields, or austere forward operating sites and can readily and routinely transition between sea bases and expeditionary airfields without loss of capability. The ACE can vary in size and composition from an aviation detachment with specific capabilities to one or more Marine Air Wings (MAW).

    d) Combat Service Support Element (CSSE). Task organized to provide the full range of CSS functions and capabilities needed to support the continued readiness and sustainability of the MAGTF as a whole. It is formed around a CSS headquarters and may vary in size and composition from a support detachment to one or more Marine Force Service Support Group (FSSG)

Marine Expeditionary Force (MEF)

A MEF is the <u>largest and most capable MAGTF</u>. Because the MEF can deploy with a formidable fighting force that can sustain itself, it is the Marine Corps' "<u>Force of Choice</u>." It is normally composed of one or more Marine divisions, Marine air wings, and Force service support groups. A Lieutenant General normally commands a MEF. <u>It comes with 60 days of sustainment</u> and the <u>CE is capable acting as a Joint/combined headquarters</u>.

1. **Permanent MEF Headquarters**
   a) I MEF—Camp Pendleton, California
   b) II MEF—Camp Le Jeune, North Carolina
   c) III MEF—Okinawa, Japan

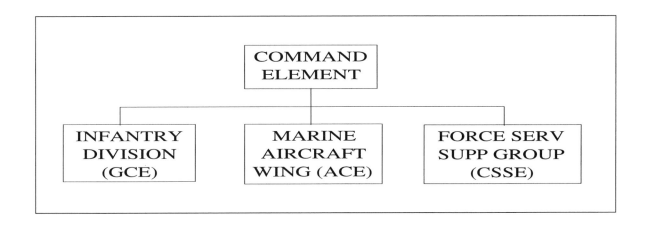

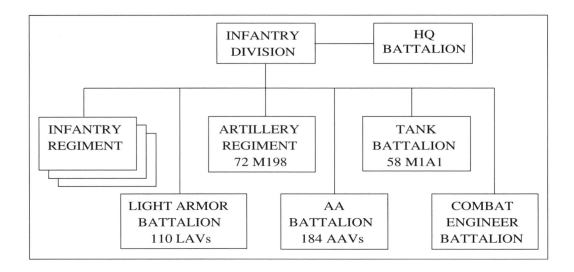

2. **Notional MEF.** A MEF's typical composition provides for the:
    a) Command Element (CE) (MEF HQ)
    b) Ground Combat Element (GCE) (MARINE DIVISION)
    c) Aviation Combat Element (ACE) (MARINE AIR WING)
    d) Combat Service Support Element (CSSE) (FORCE SERVICE SUPPORT GROUP)

Marine Division (MarDiv)

The MarDiv is the largest permanent organization of ground combat power in the Fleet Marine Force. A MarDiv may be employed as the GCE of a large landing force or provide Regimental (RLT) and/or Battalion Landing Teams (BLT) for employment with smaller landing forces. Major subordinate elements of the MarDiv are:

a) Infantry Regiment (X3)
b) Artillery Regiment
c) Tank Battalion
d) Light Armor Reconnaissance Battalion
e) Assault Amphibian Battalion
f) Combat Engineer Battalion
g) Headquarters Battalion

Marine Aircraft Wing (MAW)

The MAW is the largest organization of aviation combat power in the FMF. There are three active duty MAWs and one reserve. A MAW, which

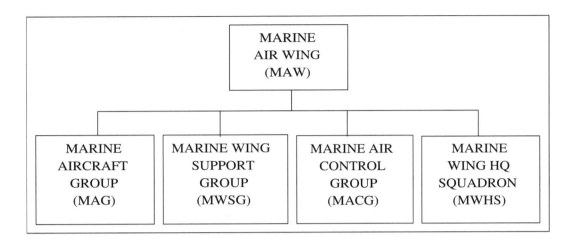

is commanded by a Major General, may be employed as the ACE of a large landing force or provide composite Marine Aircraft Groups and/or squadrons to be employed with smaller landing forces. Major subordinate elements of the MAW are:

a) Marine Air Groups (MAG)s.
    (1) Fixed wing squadrons (VMFA, VMA, VMGR).
    (2) Rotary wing squadrons (HMH, HMM, HML, HML/A).
    (3) Unmanned Aerial Vehicles (VMU).
b) Marine Wing Support Group (MWSG). Provides all essential ground support requirements to aid designated fixed-or-rotary wing components.
c) Marine Air Control Group (MACG). It's mission is to provide, operate, and maintain the Marine Air Command and Control System (MACCS). It coordinates all aspects of air command and control and air defense within the MAW.
    (1) Marine Tactical Air Command Squadron/Tactical Air Command Center (TACC)
    (2) Marine Air Control Squadron/Tactical Air Operations Center (TAOC)
    (3) Marine Air Support Squadron/Direct Air Support Center (DASC)
    (4) Low Altitude Air Defense Battalion/Battery (LAAD Bn/Btry— Stinger/60 Avengers)
    (5) Marine Wing Communications Squadron (MWCS)
d) Marine Wing Headquarters Squadron (MWHS)

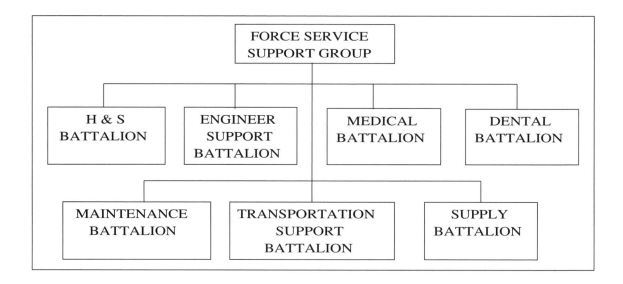

Force Service Support Group (FSSG)

The largest composite grouping of combat service support units in the FMF. There are three active duty FSSGs and one reserve FSSG. Each FSSG, which is normally commanded by a Brigadier General, may be employed as the CSSE of a large landing force or provide a task organized CSSE for employment with smaller landing forces. Major subordinate units of the FSSG are:

    a) Headquarters and Service Battalion
    b) Engineer Support Battalion
    c) Medical Battalion
    d) Dental Battalion
    e) Maintenance Battalion
    f) Transportation Support Battalion
    g) Supply Battalion

Marine Expeditionary Brigade (MEB)

The Marine Expeditionary Brigade (MEB) is the mid-sized MAGTF and is normally commanded by a brigadier general. The MEB bridges the gap between the MEU, at the tip of the spear, and the MEF, our principal war fighter. With 30 days of sufficient supplies for sustained operations, the MEB is capable of conducting amphibious assault operations and maritime prepositioning force (MPF) operations. During potential crisis situations, a MEB may be forward deployed afloat for an extended period in order to provide an immediate combat response. A MEB can operate

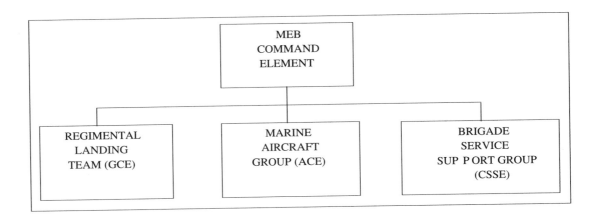

independently or serve as the advance echelon of a MEF. The MEB CE is embedded in the MEF CE and identified by line number for training and rapid deployment. The MEB can provide supported CINC's with a credible war fighting capability that is rapidly deployable and possesses the capability to impact all elements of the battlespace. If required, a MEB CE is capable of assuming the role of JTF Headquarters for small operations with additional MEF CE augmentation. As an expeditionary force, it is capable of rapid employment and employment via amphibious shipping, strategic air/sealift, geographical or maritime prepositioning force assets, or any combination thereof. There are three standing MEB command elements: 1st Marine Expeditionary Brigade, assigned within I Marine Expeditionary Force, and located at Camp Pendleton, CA; 2d Marine Expeditionary Brigade, assigned within II Marine Expeditionary Force, and located at Camp Lejeune, NC; and 3d Marine Expeditionary Brigade, assigned within III Marine Expeditionary Force, and located in Okinawa, Japan. 1st and 2d MEB CEs were activated in November 1999. 3d MEB CE was activated in January 2000.

Task Organization. The composition of a MEB varies according to the mission, forces assigned, and the area of operations. A MEB is typically organized with the following elements:

MEB Command Element. The MEB command element will provide command and control for the elements of the MEB. When missions are assigned, the notional MEB command element is tailored with required support to accomplish the mission. Detachments are assigned, as necessary, to support subordinate elements. The MEB CE is fully capable of executing all of the staff functions of a MAGTF (administration and personnel, intelligence, operations and training, logistics, plans, communications and information systems, PAO, SJA, Comptroller, and COMSEC).

Ground Combat Element (GCE). The ground combat element (GCE) is normally formed around a reinforced infantry regiment. The GCE can be composed of from two to five battalion sized maneuver elements (infantry, tanks, LAR) with a regimental headquarters, plus artillery, Assault Amphibian Bn, reconnaissance, TOWs and engineers.

Aviation Combat Element (ACE). The aviation combat element (ACE) is a composite Marine aircraft group (MAG) task-organized for the assigned mission. It usually includes both helicopters and fixed wing aircraft, and elements from the Marine wing support group and the Marine air control group. The MAG has more varied aviation capabilities than those of the aviation element of a MEU. The most significant difference is the ability to command and control aviation with the Marine Air Command and Control System (MACCS). The MAG is the smallest aviation unit designed for independent operations with no outside assistance except access to a source of supply. Each MAG is task-organized for the assigned mission and facilities from which it will operate. The ACE headquarters will be an organization built upon an augmented MAG.

Combat Service Support Element (CSSE). The brigade service support group (BSSG) is task organized to provide CSS beyond the capability of the supported air and ground elements. It is structured from personnel and equipment of the force service support group (FSSG). The BSSG provides the nucleus of the landing force support party (LFSP) and, with appropriate attachments from the GCE and ACE, has responsibility for the landing force support function when the landing force shore party group is activated.

The MEB is deployed via a continuous flow of task-organized forces building on MAGTFs. As an expeditionary force, it is capable of rapid deployment and employment via amphibious shipping, strategic air/sealift, marriage with geographical or maritime prepositioning force assets, or any combination thereof. The MEB deploys with sufficient supplies to sustain operations for 30 days. The MEB may be comprised of elements from MPF, ACF, or the ATF. Early command and control forward is critical, therefore a MEB will be deployed with enabling communications into theater as quickly as possible. The MEB provides operational agility to the MEB Commander and supports all war fighting functions: maneuver, intelligence, logistics, force protection, fires, and command headquarters or provided from other MAW assets.

Marine Expeditionary Unit (Special Operations Capable)—MEU(SOC)

In 1983, the Secretary of Defense directed each military service and defense agency to review their existing special operations capabilities and develop a plan for achieving the level of special operations capability required to combat both current and future low intensity conflicts and terrorist threats. In response, the Marine Corps instituted an aggressive SOC training program to optimize the inherent capability of MEUs to conduct selected maritime special operations.

Progressive improvement in individual and unit skills attained through enhanced training and the addition of specialized equipment allow a MEU to execute a full range of conventional and selected maritime special operations. This is accomplished by means of dedicated and intensive predeployment training program of about 26 weeks that emphasizes personnel stabilization coupled with focused, standardized, and integrated Amphibious Ready Group (ARG)/MEU training. MEUs that have undergone this enhanced training program have been provided special equipment, and have successfully completed a thorough evaluation/certification under the cognizance of the Force Commander, shall be designated as SOC. The primary goal for all MEUs shall be certification and designation as SOC prior to deployment.

The primary objective of the MEU(SOC) is to provide the National Command Authorities and geographic combatant commanders with an effective means of dealing with the uncertainties of future threats, by providing forward-deployed units which offer unique opportunities for a variety of quick reaction, sea-based, crisis response options, in either a conventional amphibious role, or in the execution of selected maritime special operations. "From the Sea" articulates the vision of MAGTF's participating in naval expeditionary forces of combined arms, which are task-organized, equipped, and trained to conduct forward presence and crisis response missions while operating in littoral areas of the world.

Organization of the Marine Expeditionary Unit (Special Operations Capable)

The forward deployed MEU (SOC) is uniquely organized and equipped to provide the naval or joint force commander with rapidly deployable, sea-based capability with 15 days of sustainment optimized for forward presence and crisis response missions. The MEU (SOC) may also serve as an enabling force for follow-on MAGTFs (or possibly joint/combined forces) in the event the situation or mission requires additional capabili-

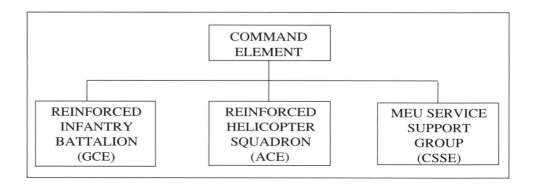

ties and resources. The MEU is comprised of a command element; a reinforced infantry battalion as the GCE; a composite helicopter squadron as the ACE; and a CSSE designated the MEU Service Support Group (MSSG). Currently there are 7 permanent MEUs. 11th, 13th, and 15th on the west coast at Camp Pendleton, CA., 22d, 24th and 26th on the east coast at Camp LeJeune and the 31st MEU in Okinawa, Japan. All MEUs have their own identical table of organization, table of equipment, and a separate monitor command code. Most importantly, there are always two deployed MEUs, two deploying MEUs and two MEUs doing the 26-week "work-up" to deployment. 31st MEU in Okinawa has recently begun participation in the regular 6-month deployment rotation.

COMMAND ELEMENT. The CE of the MEU (SOC) is a permanently established organization augmented to provide the command and control (C2) functions and the command, control, communications, computers and intelligence systems (C4I) necessary for effective planning and execution of all operations. In addition to permanently assigned Marines, the MEU CE is augmented with detachments from the MEF Headquarters Group (MHG) for deep reconnaissance, fire support, intelligence, electronic warfare, and communications.

MEU STAFF consists of Headquarters Section, Administration Section (S-1), including Staff Judge Advocate, Operations Section (S-3), Intelligence Section (S-2), Logistics Section (S-4), and Communications Section (S-6).

Maritime Special Purpose Force (MSPF). The MSPF is a unique task organization drawn from the MEU major subordinate elements. The MSPF is not designed to duplicate existing capabilities of Special Operation Forces, but is intended to focus on operations in a maritime environment. The MSPF provides the enhanced operational capability to complement or enable conventional operations or to execute special maritime operations. The MSPF cannot operate independently of its parent MEU. It relies on the MEU for logistics, intelligence, communications, transportation and supporting fires. Accordingly, command of the MSPF must remain under the control of the MEU commander. The MSPF is organized and trained to be rapidly tailored to meet a specific mission. It is notionally comprised of:

(a) Command Element. Commander, Comm. Det, Marine Liaison Group Det, Medical Section, Interrogator/Translator Team (ITT) Det and Counterintelligence (CI) Det.

(b) Covering Element. Structured around rifle platoon provided by the Battalion Landing Team and may be augmented by the Naval Special Warfare Task Unit (NSWTU). The covering element will act as a reinforcing unit, a support unit, a diversionary unit, or a extraction unit.

(c) Strike Element. Focus of effort of the MSPF and is organized to perform assault, explosive breaching, internal security, and sniper functions. FORECON Det, Security Teams, EOD Det, Combat Photo Team and possibly a NSWTU.

(d) Aviation Support Element. Provided by the ACE. Specific structure will vary, but will have the capabilities of precise night flying and navigation, plus various insertion/extraction means and forward area refueling point (FARP) operations.

(e) Reconnaissance and Surveillance Element. Normally composed of assets from the Battalion Landing Team (BLT) STA platoon (sniper support) coupled with elements of the RADBN Det, COMM Det, MLG Det, and CI/ITT assets from the MEU CE.

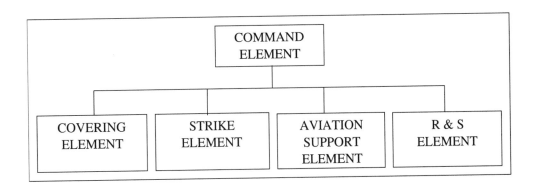

```
┌─────────────────────────────────────────────────────────────────────────┐
│                          ┌─────────────────┐                              │
│                          │    INFANTRY     │                              │
│                          │  BATTALION (+)  │                              │
│                          └─────────────────┘                             │
│                                                                           │
│  ┌───────────┐   ┌───────────┐    ┌───────────┐   ┌───────────┐          │
│  │ INFANTRY  │   │  WEAPONS  │    │ ARTILLERY │   │   LIGHT   │           │
│  │  COMPANY  │   │  COMPANY  │    │  BATTERY  │   │  ARMORED  │           │
│  └───────────┘   └───────────┘    └───────────┘   │   RECON   │          │
│                                                    └───────────┘          │
│  ┌───────────┐   ┌───────────┐    ┌───────────┐   ┌───────────┐          │
│  │  ASSAULT  │   │   TANK    │    │    TOW    │   │  COMBAT   │           │
│  │ AMPHIBIAN │   │  PLATOON  │    │  SECTION  │   │ ENGINEER  │           │
│  │  PLATOON  │   └───────────┘    └───────────┘   │  PLATOON  │          │
│  └───────────┘                                    └───────────┘          │
│           ┌────────────┐   ┌────────────┐   ┌────────────┐               │
│           │ SURV & TGT │   │ SHORE FIRE │   │   RECON    │               │
│           │ACQUISITION │   │  CONTROL   │   │  PLATOON   │               │
│           │  PLATOON   │   │   PARTY    │   └────────────┘               │
│           └────────────┘   └────────────┘                               │
└─────────────────────────────────────────────────────────────────────────┘
```

Ground Combat Element (GCE)

The GCE is normally structured around a reinforced infantry battalion that forms a BATTALION LANDING TEAM (BLT). Specific reinforcements will vary, but generally include artillery, reconnaissance, light armor (maybe tanks), anti-armor, amphibious assault vehicles, and combat engineer attachments. The battalion consists of an H&S company, three letter companies, and a weapons company. There are two important things to note about the BLT, first, unlike a standard infantry battalion, the BLT (when formed) comes to full strength in personnel and equipment (T/O & T/E). Secondly, the companies within the BLT have become specialized. One company in the BLT specializes in Mechanized operations, one company specializes in Helo Operations, while the third company specializes in small boat operations.

Aviation Combat Element (ACE)

The ACE is a reinforced helicopter squadron that includes AV-8B Harrier attack aircraft, and two CONUS based KC-130 aircraft. The ACE is task organized to provide assault support, fixed wing and rotary wing close air support, airborne command and control, and low-level, close-in air defense. The ACE is structured as follows:

(a) HMM SQUADRON: Marine Medium Helicopter Squadron configured with twelve CH-6E helicopters: provides medium-lift assault support.

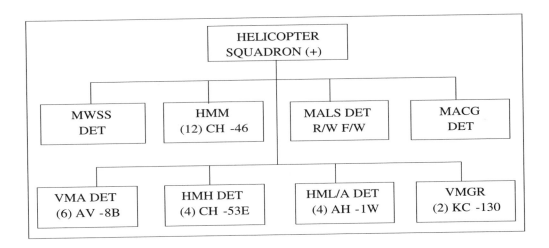

(b) HMH DET: Marine Heavy Helicopter Squadron detachment configured with CH-53E helicopters: provides extended range, heavy-lift assault.

(c) HML/A DET: Marine Light Attack Squadron detachment configured with four AH-1W attack helicopters, and three UN-1N utility helicopters: provides close air support, airborne command and control, and escort.

(d) VMA DET: Marine Attack Squadron detachment configured with six AV-8B Harrier aircraft: provides organic close air support. When appropriate shipping (i.e., LHA, LHD) is not available, the detachment trains with the MEU throughout predeployment training, and then is placed on CONUS standby and prepared to deploy within 96 hours.

(e) VMGR DET: Marine Aerial Refueler/Transport Squadron detachment configured with two KC-130 aircraft: provides refueling services for embarked helicopters and AV-8B aircraft, and performs other tasks (i.e., parachute operations, flare drops, cargo transportation, etc.) as required. The detachment trains with the MEU throughout predeployment training, and then is placed on CONUS standby and prepared to deploy within 96 hours.

(f) MARINE AIR CONTROL GROUP DET

    i) LOW ALTITUDE AIR DEFENSE (LAAD) BATTALION DET: Provides low level, close-in air defense utilizing MAN-PAD and the Avenger Stinger Missile Systems.

    ii) MARINE AIR SUPPORT SQUADRON DET.: Provides a limited Direct Air Support Center (DASC) capability for enhanced integration of air support into the MEU(SOC) scheme of maneuver.

(g) MARINE WING SUPPORT SQUADRON DET: Provides aviation bulk fuel and limited food service support.

(h) MARINE AVIATION LOGISTICS SQUADRON DET: Provides intermediate maintenance and aviation supply support.

Combat Service Support Element (CSSE):
The CSSE is a MEU Service Support Group (MSSG) which provides the full range of combat service support including supply, maintenance, transportation, deliberate engineering, medical and dental, automated information processing, utilities, landing support (port/airfield support operations), disbursing, legal, and postal services and 15 days of sustainability (Class I, II, III (B), IV, V, VIII, IX) necessary to support MEU (SOC) assigned missions.

Missions of a MEU(SOC): The MEU(SOC) is a self-sustained, amphibious, combined arms air-ground task force capable of conventional and selected maritime special operations of limited duration in support of a Combatant commander. The following is the mission statement from MCO 3120.9A:

To provide the geographic combatant commander a forward-deployed, rapid crisis response capability by conducting conventional amphibious and selected maritime special operations under the following conditions: at night; under adverse weather conditions; from over the horizon; under emissions control; from the sea, by surface and/or by air; commence execution within 6 hours of receipt of the warning order. To act as an enabling force for a follow-on MAGTF or joint and/or combined forces in support in support of various contingency requirements.

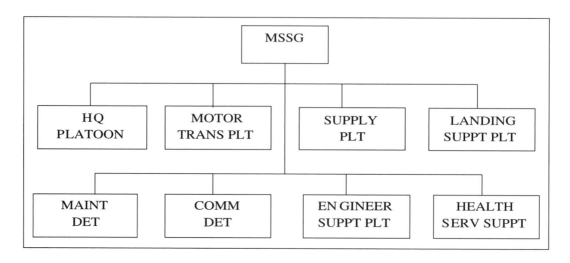

The inherent capabilities of a forward-deployed MEU(SOC) are divided into four broad categories:

(1) Amphibious Operations
    (a) Amphibious Assault
    (b) Amphibious Raid
    (c) Amphibious Demonstration
    (d) Amphibious Withdrawal
(2) Direct Action Operations
    (a) In-Extremis Hostage Recovery (IHR)
    (b) Seizure/Recovery of Offshore Energy Facilities.
    (c) Visit, Board, Search and Seizure Operations (VBSS)
    (d) Specialized Demolition Operations
    (e) Tactical Recovery of Aircraft and Personnel (TRAP)
    (f) Seizure/Recovery of Selected Personnel or Material
    (g) Counterproliferation (CP) of Weapons of Mass Destruction (WMD)
(3) Military Operations Other Than War (MOOTW)
    (a) Peace Operations
        i) Peacekeeping
        ii) Peace Enforcement
    (b) Security Operations
        i) Non-combatant Evacuation Operations (NEO)
        ii) Reinforcement Operations
    (c) Joint/Combined Training/Instruction Team
    (d) Humanitarian Assistance/Disaster relief
(4) Supporting Operations
    (a) Tactical Deception Operations
    (b) Fire Support Planning, Coordination, and Control in a Joint/Combined Environment
    (c) Signal Intelligence (SIGINT)/Electronic Warfare (EW)
    (d) Military Operations in Urban Terrain (MOUT)
    (e) Reconnaissance and Surveillance (R&S)
    (f) Initial Terminal Guidance (ITG)
    (g) Counterintelligence Operations (CI)
    (h) Airfield/Port Seizure
    (i) Limited Expeditionary Airfield Operations
    (j) Show of Force Operations
    (k) JTF Enabling Operations
    (l) Sniping Operations

The MEU(SOC) has a limited:

(1) Defensive capability against armored/motorized units in open terrain.
(2) Defensive capability against a sustained low-level air attack when operating independent of naval air support.
(3) Capability to replace combat losses and retrain if early introduction of follow-on forces is not contemplated.
(4) Capability to participate in special warfare tasks requiring mobile training teams or nation-building efforts. However, the MEU(SOC) can provide some entry level and/or reinforcement training.
(5) Ability to establish a MEU Headquarters ashore, and operate independent of Naval Shipping. The MEU(SOC) is heavily reliant upon shipboard facilities for C4I and aviation maintenance support.

The most significant difference between the current MEU (SOC) program and the old training for Marine Amphibious Units (MAU) in the 1970s/early 1980's is that an intense 26-week "work-up" exists. The training program is standardized and follows a progressive building-block approach to training. This training program integrates the Amphibious Squadron (PHIBRON) and the MEU as well as other designated forces (i.e. CVBG) to optimize coordination and use of capabilities. The 26-week "work-up" culminates in a Special Operations Capable Exercise (SOCEX) that realistically evaluates the Mau's war fighting capabilities. Only MEUs which have demonstrated proficiency in the skills and capabilities listed above will be designated as "MEU(SOC)."

Special Purpose MAGTF (SPMAGTF)

A special-purpose MAGTF (SPMAGTF) is a non-standing MAGTF temporarily formed to conduct a specific mission. It is normally formed when a standing MAGTF is either inappropriate or unavailable. SPMAGTFs are organized, trained, and equipped to conduct a wide variety of missions ranging from crisis response, to regionally focused training exercises, to peacetime missions. Their SPMAGTF designation derives from the mission they are assigned, the location in which they will operate, or the name of the exercise in which they will participate (e.g., "SPMAGTF (X)," "SPMAGTF Somalia," " SPMAGTF UNITAS," "SPMAGTF Andrew," etc.).

Estimate of the Situation

Commander's Estimate

The commander's estimate of the situation is an orderly reasoning process by which a commander evaluates all significant factors affecting the situation. At higher levels of command, the commander may make his estimate based on the written or verbal estimates his staff develops. At these levels, it is customary for staff officers to **assist** the commander by:

- Assembling and interpreting information
- Making necessary assumptions
- Formulating possible courses of action
- Analyzing and comparing possible courses of action

METT-T

The commander uses the process represented by the acronym "METT-T" to analyze the situation. Each letter of the acronym stands for an important aspect of the situation.

ACRONYM	ELEMENT OF
M	MISSION
E	ENEMY
T	TERRAIN/WEATHER
T	TROOPS & FIRE SUPPORT AVAILABLE
T	TIME AVAILABLE

Logistics and space: METT-T

Commanders also consider the impact of logistics and any logistical requirements as well as the advantages and limitations of space.

> **NOTE:** *Logistics* **is usually considered as part of "troops and fire support available" (friendly combat power), while various aspects of** *space* **are considered under both "terrain" (maneuver area available) and time (the time/space relationship). Sometimes the acronym METT-TSL is used: Mission, Enemy, Terrain and Weather, Troops and Fire Support Available-Time-Space and Logistics.**

Mission

The estimate of the situation is a systematic approach to problem solving. It helps the commander examine all relevant factors that affect his mission as well as select a feasible course of action that will accomplish his assigned responsibilities. The commander uses the acronym **METT-T** to analyze the mission and make his estimate.

Analyzing your mission is one of five factors in the estimate of the situation.

- The mission is the **task,** together with its **purpose,** which clearly indicates the reason for the mission and the action to be taken.
- The mission must be carefully analyzed and understood within the context of the higher commander's intent. Specified and implied tasks are derived and assigned a relative priority.
- A mission clearly conveys the WHO, WHAT, WHEN, WHERE, and WHY of an operation. For example:
  "AT H-Hour on D-Day, BLT 1/5 attacks toward Objective A in order to fix the enemy in the vicinity of Hill 138."

Who: BLT 1/5
What: attacks
When: H-Hour on D-Day
Where: Objective A
Why: . . . in order to fix the enemy

Enemy

The second factor to consider is the enemy situation. The commander considers all of the enemy's capabilities. He makes plans to counteract or neutralize those enemy capabilities that can prevent or hinder accomplishing his mission.

As the commander considers the enemy situation, he carefully distinguishes what he knows about the enemy from what he guesses or assumes about the enemy. He studies **known** dispositions, compositions, and strengths of the enemy in terms of committed, reinforcing, and supporting forces. He attempts to "see" the enemy on the terrain. He mentally reviews recent significant activities, and he strives to find enemy weaknesses and peculiarities that can be exploited.

What are enemy capabilities? Enemy capabilities are:

"Those courses of action of which the enemy is physically capable, and that, if adopted, will affect accomplishment of our mission. The term 'capabilities' includes not only general courses of action open to the enemy, such as attack, defense, or withdrawal (DRAWD), but also all the particular courses of action possible under each general course of action. Enemy capabilities are considered in the light of all known factors affecting military operations, including time, space, weather, terrain, and the strength and disposition of enemy forces."

Determining the enemy commander's capabilities will become the key to any analysis of the enemy. It is not good enough to know numerous facts about the enemy unless you are capable of drawing conclusion as to his next action. It is only through this process that you are able to outmaneuver our opponent by anticipating his moves.

The commander determines what information he lacks. The successful commander is especially careful not to make arrogant assumptions about the enemy capabilities base upon race, religion, nationality, literacy, sophistication, sex, etc. Time and time again throughout history, military forces have suffered disastrous defeats because of overconfidence brought on by some false assumption of natural or national superiority.

Two important acronyms are used to communicate the enemy's strength, composition, weapons, combat efficiency, and capabilities. The acronyms SALUTE and DRAWD are the basis for further analysis of enemy capabilities. They are **tools** to assist in the evaluation and analysis of the enemy.

Use **SALUTE** to evaluate the enemy's known situation:

Size	Unit
Activity	Time
Location	Equipment

Use **DRAWD** to evaluate enemy capabilities. Can the enemy:

Defend	Withdraw
Reinforce	Delay
Attack	

Troops and Fire Support Available

When commanders consider troops and the available fire support, they are developing their assessment of their **combat power.** Joint Publication 1-

02 defines combat power as *"the total means of destructive and/or disruptive force which a military unit/formation can apply against the opponent at a given time."*

Let's look at this definition:

- First of all, combat power is **always** relative. It is not an absolute. In this respect, combat power is much like maneuver in that its value has meaning only when considered in relation to the enemy's combat power.
- Secondly, combat power is composed of both tangible and intangible factors. You must avoid the tendency to equate combat power with "bean counting" or "number crunching." For example, recall the old three-to-one advantage that the attacker should have to the defender: If the enemy has 200 men defending a position, you must have at least 600 troops to seize that position. As a former Commandant of the Marine Corps General Barrow said, *"Success in battle is not a function of how many show up, but who they are."*

If you use your imagination to generate superior combat power at a decisive time and place and are aggressive in employing it; you will win against otherwise seemingly very high odds.

This fourth factor of METT-T includes considerations of friendly capabilities. Some of the capabilities the commander will normally consider are:

- equipment and weapons
  - tanks, infantry fighting vehicles, armored personnel carriers
  - machine guns
  - Anti-tank missiles and mortars
- air support available
- artillery support available
- logistics capabilities
- morale
- training
- leadership
- Fire Support Available
  - There are many assets for consideration when talking about fire support available as is mentioned above. Combined arms is the full integration of arms in such a way that in order to counteract one, the enemy must make himself more vulnerable to another. We pose the enemy not just with a problem, but with a dilemma—a no win situation. The Marine Corps takes advantage of the complementary

characteristics of different types of units and enhances its mobility and firepower. For example, in order to defend against the infantry attack, the enemy must make himself vulnerable to the supporting arms. We use assault support to quickly concentrate superior ground forces for a breakthrough. We use artillery and close air support to support to support the infantry penetration, and we use deep air support to interdict enemy reinforcements. The plan of attack must consist of a scheme of maneuver and a fire support plan. These two aspects must complement each other. For example, at the very lowest level is the complementary use of the automatic weapon and grenade launcher within a fire team. We pin an enemy down with the high-volume, direct fire of the automatic weapon, making him a vulnerable target for the grenade launcher. If he moves to escape the impact of the grenades, we engage him with the automatic weapon.

- Troops
  - Within the "Troops and Fire Support Available" portion of the estimate process, you must consider logistics. Joint Publication 1-02 defines logistics as *"the science of planning and carrying out the movement and maintenance of forces."*
  - One description of the role logistics plays in the commander's estimate comes from British General Archibald Wavell:

"The more I see of war, the more I realize how it all depends on administration and transportation . . . It takes little skill or imagination to see <u>where</u> you would like your army to be and <u>when</u>; it takes much knowledge and hard work to know where you can place your forces and whether you can maintain them there."

  - A real knowledge of supply and movement factors must be the basis of every leader's plan. Only then can the leader know how and when to take risks with those factors. The key point in the quotation above is:
    "it takes much knowledge and hard work to know where you can place your forces and whether you can maintain them there."

The following historical case studies are good examples of the importance of conducting an estimate of the situation.

Terrain and Weather

The fourth factor of the estimate of the situation is to analyze the terrain that you will operate in and the weather conditions.

COKOA is a mnemonic device used for identifying terrain information which the leader derives from a personal reconnaissance or through a careful map study. COKOA stands for:

Cover and concealment
Obstacles
Key terrain
Observation and fields of fire
Avenues of approach

An **obstacle** is anything, including a natural or artificial terrain feature, that stops, impedes, or diverts military movement. Obstacles may either be existing or reinforcing.

- The mission influences the determination of obstacles. In the attack, the commander considers the features within his unit's zone of action. In the commander's opinion, an obstacle may be an advantage or disadvantage.
- Obstacles **perpendicular** to the direction of attack favor the defender by slowing or canalizing the attacker. Obstacles **parallel** to the direction of attack may assist in protecting a flank of the attacking force.
- To fully understand what constitutes an obstacle, commanders must first consider their mission and means of mobility.

An **avenue of approach** is: *"An air or ground route of an attacking force of a given size leading to its objective or to key terrain in its path."*

What is really meant by an avenue of approach? An avenue of approach is a route by which a force may reach its objective. In planning for use of avenues of approach you should consider:

- All of the other factors of key terrain, observation and fire, concealment and cover, and obstacles, from both friendly and enemy perspectives.
- The mission, type, and size of the unit.

Generally, commanders must consider avenues of approach that are adequate for the forces operating in the maneuver space. Intelligence about each avenue is checked against possible schemes of maneuver, both friendly and enemy.—Maneuver space should be adequate. This consideration is based on deployment patterns and the means of movement. Ease of movement is considered and includes such factors as:

- soil type
- river trafficability
- terrain compartments
- road trafficability
- steepness of slopes
- vegetation

A **key terrain feature** is: "*Any locality or area the seizure or retention of which affords a marked advantage to either combatant.*"

Based on the mission of the command, commanders must consider key terrain features in formulating courses of action.

Terrain which **permits** or **denies** maneuver may be key terrain.

Tactical use of terrain often emphasizes increasing the ability to apply friendly combat power at the same time forcing the enemy into areas which reduce his combat power. Terrain which permits this may also by key terrain. Considerations in selecting key terrain are:

- the effect of terrain on fire and maneuver;
- the application of combat power;
- the preservation of force integrity.

A terrain feature may afford a marked advantage in one set of circumstances but little or no advantage under other conditions. Selecting key terrain varies with the level of command, the type of unit, and the mission of the unit.

Combat service support (CSS) and aviation units need key terrain:

- CSS units need roads over which to move supplies and secure areas in which to establish facilities.
- Aviation units need high terrain on which to set up radar and communication facilities and large flat areas for airfields.

Observation and fields of fire are so closely related that they are considered together. They are not synonymous, but fields of fire are based on observation because troops must see a target to bring effective fire upon it.

"Observation is the area over which surveillance can be exercised either visually or through the use of surveillance devices, both optical and electronic."

A field of fire is: *"The area a weapon or group of weapons may cover effectively with fire from a given position."*

Observation varies with
- weather conditions
- the time of day
- vegetation
- friendly and enemy smoke
- surrounding terrain

Observation generally is best form the highest terrain features. However, during times of poor visibility, positions in low areas that the enemy must pass through may provide better observation than high points form which nothing can be seen.

Fields of fire for direct fire weapons, such as machine guns and automatic rifles, may be affected by terrain conditions between the weapon and the target.

A leader identifies those terrain features within the area of operations and those areas adjacent to the area of operations. These are terrain features which afford the friendly or enemy force favorable observation and fire.

Concealment and cover are protection from observation and fire. Specifically, concealment is *"the protection form observation or surveillance."*

"Cover is protection from the effects of fire."

What is good concealment? Woods, underbrush, snowdrifts, tall grasses, cultivated vegetation, or any other feature which denies observation usually provides good concealment. Good concealment may also be provided by weather conditions, such as fog and rain, and by darkness. Concealment from ground observation does not necessarily provide concealment from air observation or from electronic or infrared detection devices.

Remember, terrain that provides concealment **may** or **may not** provide cover!

Cover may be provided by rocks, ditches, quarries, caves, river banks, folds in the ground, shell craters, buildings, walls, railroad embankments and cuts, sunken roads, and highway fills. Areas that provide cover from direct fires may or may not protect against the effects of indirect fire. Most terrain features that offer cover also provide concealment from ground observation.

Although you have studied the military aspects of terrain and weather separately, terrain and weather are, in fact, inseparable. For example, terrain that offers good trafficability when it is dry may be impassable when it is wet. A hill that offers good observation on a clear day may not provide any visibility on a rainy day or during foggy conditions. Now, let's turn to our next factor in the estimate of the situation.

"Weather is the state of the atmosphere at a given time and place. It includes atmospheric pressure, winds, humidity, clouds and fog, precipitation, and fronts or zones where air masses of different temperatures meet."

What does all that mean to a military force? Let's consider this quote from General Eisenhower's book, Crusade in Europe:

"Some soldier once said,, 'The weather is always neutral.' Nothing could be more untrue. Bad weather is obviously the enemy of the side that seeks to launch projects requiring good weather, or of the side possessing great assets, such as strong air forces, which depend upon good weather for effective operation."

Eisenhower means that a certain type of weather in and of itself is not always good or bad. Whether the weather is good or bad for an operation depends on the requirements of a specific operation, conducted in response to a specific situation. The key point is that what U.S. forces traditionally refer to as "bad weather" can be a great ally.

Time

Time is the commander's most important resource! What is meant by the term time available? Perhaps the best and clearest comment about the role of time in the offense is Clausewitz's injunction: *"Time which is allowed to pass unused accumulates to the credit of the defender."*

Time is closely tied to the concepts of **momentum** and **culminating point.** If the attacker is to succeed, he must constantly concentrate on imaginatively and aggressively getting the most from each moment. Time is closely linked to space.

Make a time plan. The goal is to give the subordinate unit enough daylight to conduct planning, reconnaissance, and preparation before the start of combat operations. It does more harm than good to present a perfect plan to subordinate units if they do not have the time to disseminate their own orders and prepare. Plan your time by using the "half-rule" or the "one-third, two-thirds rule."(For example, half of the available time goes to the commander and half goes to the subordinate units.)

Backward planning is also necessary to ensure an effective use of time. Start with the last known action and progress backwards to present time. This will be the time for crossing of the line of departure for an offensive battle or from the time the defense must be established.

The commander must consider the distance he must move in the required time. This is why time and space are considered together. The commander should compute:

- how much time will be needed to move certain distances;
- how far form the objective he must begin to change formations to begin the assault.

Compute this time/distance estimate with regards to specific conditions, such as weather or the enemy situation. You must anticipate friction, such as obstacles or harassing fire from the enemy that your troops may encounter. It will slow down friendly units.

Review Questions

1. What is the goal of Maneuver Warfare?
2. What is the basis for the Marine Corps Philosophy of Command?
3. What are the three phases of the offense?
4. Just before the Final Coordination Line the assault element is in what position?
5. Define: METT-T
        DRAW-D
        COKOA
        SALUTE
6. What is combined arms?

Suggested Further Reading

MCDP 1: *Warfighting,* Department of the Navy, Headquarters Marine Corps, Washington, D.C, 1989.

MCDP 1-3: Tactics, Department of the Navy, Headquarters Marine Corps, Washington, D.C, 1997.

11

National Security Strategy

The great struggles of the twentieth century between liberty and totalitarianism ended with a decisive victory for the forces of freedom—and a single sustainable model for national success: freedom, democracy, and free enterprise. In the twenty-first century, only nations that share a commitment to protecting basic human rights and guaranteeing political and economic freedom will be able to unleash the potential of their people and assure their future prosperity. People everywhere want to be able to speak freely; choose who will govern them; worship as they please; educate their children—male and female; own property; and enjoy the benefits of their labor. These values of freedom are right and true for every person, in every society—and the duty of protecting these values against their enemies is the common calling of freedom-loving people across the globe and across the ages.

Today, the United States enjoys a position of unparalleled military strength and great economic and political influence. In keeping with our heritage and principles, we do not use our strength to press for unilateral advantage. We seek instead to create a balance of power that favors human freedom: conditions in which all nations and all societies can choose for themselves the rewards and challenges of political and economic liberty. In a world that is safe, people will be able to make their own lives better. We will defend the peace by fighting terrorists and tyrants. We will preserve the peace by building good relations among the great powers. We will extend the peace by encouraging free and open societies on every continent.

Defending our Nation against its enemies is the first and fundamental commitment of the Federal Government. Today, that task has changed dramatically. Enemies in the past needed great armies and great industrial capabilities to endanger America. Now, shadowy networks of individuals can bring great chaos and suffering to our shores for less than it costs to purchase a single tank. Terrorists are organized to penetrate open societies and to turn the power of modern technologies against us.

To defeat this threat we must make use of every tool in our arsenal—military power, better homeland defenses, law enforcement, intelligence, and vigorous efforts to cut off terrorist financing. The war against terrorists of global reach is a global enterprise of uncertain duration. America will help nations that need our assistance in combating terror. And America will hold to account nations that are compromised by terror, including those who harbor terrorists—because the allies of terror are the enemies of civilization. The United States and countries cooperating with us must not allow the terrorists to develop new home bases. Together, we will seek to deny them sanctuary at every turn.

The gravest danger our Nation faces lies at the crossroads of radicalism and technology. Our enemies have openly declared that they are seeking weapons of mass destruction, and evidence indicates that they are doing so with determination. The United States will not allow these efforts to succeed. We will build defenses against ballistic missiles and other means of delivery. We will cooperate with other nations to deny, contain, and curtail our enemies' efforts to acquire dangerous technologies. And, as a matter of common sense and self-defense, America will act against such emerging threats before they are fully formed. We cannot defend America and our friends by hoping for the best. So we must be prepared to defeat our enemies' plans, using the best intelligence and proceeding with deliberation. History will judge harshly those who saw this coming danger but failed to act. In the new world we have entered, the only path to peace and security is the path of action.

As we defend the peace, we will also take advantage of an historic opportunity to preserve the peace. Today, the international community has the best chance since the rise of the nation-state in the seventeenth century to build a world where great powers compete in peace instead of continually prepare for war. Today, the world's great powers find ourselves on the same side—united by common dangers of terrorist violence and chaos. The United States will build on these common interests to promote global security. We are also increasingly united by common values. Russia is in the midst of a hopeful transition, reaching for its democratic future and a partner in the war on terror. Chinese leaders are discovering that economic freedom is the only source of national wealth. In time, they will find that social and political freedom is the only source of national greatness. America will encourage the advancement of democracy and economic openness in both nations, because these are the best foundations for domestic stability and international order. We will strongly resist aggression from other great powers—even as we welcome their peaceful pursuit of prosperity, trade, and cultural advancement.

Finally, the United States will use this moment of opportunity to extend the benefits of freedom across the globe. We will actively work to bring the hope of democracy, development, free markets, and free trade to every corner of the world. The events of September 11, 2001, taught us that weak states, like Afghanistan, can pose as great a danger to our national interests as strong states. Poverty does not make poor people into terrorists and murderers. Yet poverty, weak institutions, and corruption can make weak states vulnerable to terrorist networks and drug cartels within their borders.

The United States will stand beside any nation determined to build a better future by seeking the rewards of liberty for its people. Free trade and free markets have proven their ability to lift whole societies out of poverty—so the United States will work with individual nations, entire regions, and the entire global trading community to build a world that trades in freedom and therefore grows in prosperity. The United States will deliver greater development assistance through the New Millennium Challenge Account to nations that govern justly, invest in their people, and encourage economic freedom. We will also continue to lead the world in efforts to reduce the terrible toll of HIV/AIDS and other infectious diseases.

In building a balance of power that favors freedom, the United States is guided by the conviction that all nations have important responsibilities. Nations that enjoy freedom must actively fight terror. Nations that depend on international stability must help prevent the spread of weapons of mass destruction. Nations that seek international aid must govern themselves wisely, so that aid is well spent. For freedom to thrive, accountability must be expected and required.

We are also guided by the conviction that no nation can build a safer, better world alone. Alliances and multilateral institutions can multiply the strength of freedom-loving nations. The United States is committed to lasting institutions like the United Nations, the World Trade Organization, the Organization of American States, and NATO as well as other long-standing alliances. Coalitions of the willing can augment these permanent institutions. In all cases, international obligations are to be taken seriously. They are not to be undertaken symbolically to rally support for an ideal without furthering its attainment.

Freedom is the non-negotiable demand of human dignity; the birthright of every person—in every civilization. Throughout history, freedom has been threatened by war and terror; it has been challenged by the clashing wills of powerful states and the evil designs of tyrants; and it has been tested by widespread poverty and disease. Today, humanity holds in its hands the opportunity to further freedom's triumph over all these foes. The United States welcomes our responsibility to lead in this great mission.

George W. Bush
THE WHITE HOUSE,
September 17, 2002

I. Overview of America's International Strategy

"Our Nation's cause has always been larger than our Nation's defense. We fight, as we always fight, for a just peace—a peace that favors liberty. We will defend the peace against the threats from terrorists and tyrants. We will preserve the peace by building good relations among the great powers. And we will extend the peace by encouraging free and open societies on every continent."

President Bush
West Point, New York
June 1, 2002

The United States possesses unprecedented—and unequaled-—strength and influence in the world. Sustained by faith in the principles of liberty, and the value of a free society, this position comes with unparalleled responsibilities, obligations, and opportunity. The great strength of this nation must be used to promote a balance of power that favors freedom.

For most of the twentieth century, the world was divided by a great struggle over ideas: destructive totalitarian visions versus freedom and equality.

That great struggle is over. The militant visions of class, nation, and race which promised utopia and delivered misery have been defeated and discredited. America is now threatened less by conquering states than we are by failing ones. We are menaced less by fleets and armies than by catastrophic technologies in the hands of the embittered few. We must defeat these threats to our Nation, allies, and friends.

This is also a time of opportunity for America. We will work to translate this moment of influence into decades of peace, prosperity, and liberty. The U.S. national security strategy will be based on a distinctly American internationalism that reflects the union of our values and our national interests. The aim of this strategy is to help make the world not just safer but better. Our goals on the path to progress are clear: political and economic freedom, peaceful relations with other states, and respect for human dignity.

And this path is not America's alone. It is open to all. To achieve these goals, the United States will:

- champion aspirations for human dignity;
- strengthen alliances to defeat global terrorism and work to prevent attacks against us and our friends;
- work with others to defuse regional conflicts;
- prevent our enemies from threatening us, our allies, and our friends, with weapons of mass destruction;
- ignite a new era of global economic growth through free markets and free trade;
- expand the circle of development by opening societies and building the infrastructure of democracy;
- develop agendas for cooperative action with other main centers of global power; and
- transform America's national security institutions to meet the challenges and opportunities of the twenty-first century.

II. Champion Aspirations for Human Dignity

"Some worry that it is somehow undiplomatic or impolite to speak the language of right and wrong. I disagree. Different circumstances require different methods, but not different moralities."

President Bush
West Point, New York
June 1, 2002

In pursuit of our goals, our first imperative is to clarify what we stand for: the United States must defend liberty and justice because these principles are right and true for all people everywhere. No nation owns these aspirations, and no nation is exempt from them. Fathers and mothers in all societies want their children to be educated and to live free from poverty and violence. No people on earth yearn to be oppressed, aspire to servitude, or eagerly await the midnight knock of the secret police.

America must stand firmly for the nonnegotiable demands of human dignity: the rule of law; limits on the absolute power of the state; free speech; freedom of worship; equal justice; respect for women; religious and ethnic tolerance; and respect for private property.

These demands can be met in many ways. America's constitution has served us well. Many other nations, with different histories and cultures, facing different circumstances, have successfully incorporated these core principles into their own systems of governance. History has not been kind to those nations which ignored or flouted the rights and aspirations of their people.

America's experience as a great multi-ethnic democracy affirms our conviction that people of many heritages and faiths can live and prosper in peace. Our own history is a long struggle to live up to our ideals. But even in our worst moments, the principles enshrined in the Declaration of Independence were there to guide us. As a result, America is not just a stronger, but is a freer and more just society.

Today, these ideals are a lifeline to lonely defenders of liberty. And when openings arrive, we can encourage change—as we did in central and eastern Europe between 1989 and 1991, or in Belgrade in 2000. When we see democratic processes take hold among our friends in Taiwan or in the Republic of Korea, and see elected leaders replace generals in Latin America and Africa, we see examples of how authoritarian systems can evolve, marrying local history and traditions with the principles we all cherish.

Embodying lessons from our past and using the opportunity we have today, the national security strategy of the United States must start from these core beliefs and look outward for possibilities to expand liberty.

Our principles will guide our government's decisions about international cooperation, the character of our foreign assistance, and the allocation of resources. They will guide our actions and our words in international bodies.

We will:

- speak out honestly about violations of the nonnegotiable demands of human dignity using our voice and vote in international institutions to advance freedom;

- use our foreign aid to promote freedom and support those who struggle non-violently for it, ensuring that nations moving toward democracy are rewarded for the steps they take;
- make freedom and the development of democratic institutions key themes in our bilateral relations, seeking solidarity and cooperation from other democracies while we press governments that deny human rights to move toward a better future; and
- take special efforts to promote freedom of religion and conscience and defend it from encroachment by repressive governments.

We will champion the cause of human dignity and oppose those who resist it.

III. Strengthen Alliances to Defeat Global Terrorism and Work to Prevent Attacks Against Us and Our Friends

> *"Just three days removed from these events, Americans do not yet have the distance of history. But our responsibility to history is already clear: to answer these attacks and rid the world of evil. War has been waged against us by stealth and deceit and murder. This nation is peaceful, but fierce when stirred to anger. The conflict was begun on the timing and terms of others. It will end in a way, and at an hour, of our choosing."*

President Bush
Washington, D.C. (The National Cathedral)
September 14, 2001

The United States of America is fighting a war against terrorists of global reach. The enemy is not a single political regime or person or religion or ideology. The enemy is terrorism—premeditated, politically motivated violence perpetrated against innocents.

In many regions, legitimate grievances prevent the emergence of a lasting peace. Such grievances deserve to be, and must be, addressed within a political process. But no cause justifies terror. The United States will make no concessions to terrorist demands and strike no deals with them. We make no distinction between terrorists and those who knowingly harbor or provide aid to them.

The struggle against global terrorism is different from any other war in our history. It will be fought on many fronts against a particularly elusive enemy over an extended period of time. Progress will come through the persistent accumulation of successes—some seen, some unseen.

Today our enemies have seen the results of what civilized nations can, and will, do against regimes that harbor, support, and use terrorism to achieve their political goals. Afghanistan has been liberated; coalition forces continue to hunt down the Taliban and al-Qaida. But it is not only this battlefield on which we will engage terrorists. Thousands of trained terrorists remain at large with cells in North America, South America, Europe, Africa, the Middle East, and across Asia.

Our priority will be first to disrupt and destroy terrorist organizations of global reach and attack their leadership; command, control, and communications; material support; and finances. This will have a disabling effect upon the terrorists' ability to plan and operate.

We will continue to encourage our regional partners to take up a coordinated effort that isolates the terrorists. Once the regional campaign localizes the threat to a particular state, we will help ensure the state has the military, law enforcement, political, and financial tools necessary to finish the task.

The United States will continue to work with our allies to disrupt the financing of terrorism. We will identify and block the sources of funding for terrorism, freeze the assets of terrorists and those who support them, deny terrorists access to the international financial system, protect legitimate charities from being abused by terrorists, and prevent the movement of terrorists' assets through alternative financial networks.

However, this campaign need not be sequential to be effective, the cumulative effect across all regions will help achieve the results we seek. We will disrupt and destroy terrorist organizations by:

- direct and continuous action using all the elements of national and international power. Our immediate focus will be those terrorist organizations of global reach and any terrorist or state sponsor of terrorism which attempts to gain or use weapons of mass destruction (WMD) or their precursors;
- defending the United States, the American people, and our interests at home and abroad by identifying and destroying the threat before it reaches our borders. While the United States will constantly strive to enlist the support of the international community, we will not hesitate to act alone, if necessary, to exercise our right of self-defense by acting preemptively against such terrorists, to prevent them from doing harm against our people and our country; and

- denying further sponsorship, support, and sanctuary to terrorists by convincing or compelling states to accept their sovereign responsibilities. We will also wage a war of ideas to win the battle against international terrorism. This includes:
- using the full influence of the United States, and working closely with allies and friends, to make clear that all acts of terrorism are illegitimate so that terrorism will be viewed in the same light as slavery, piracy, or genocide: behavior that no respectable government can condone or support and all must oppose;
- supporting moderate and modern government, especially in the Muslim world, to ensure that the conditions and ideologies that promote terrorism do not find fertile ground in any nation;
- diminishing the underlying conditions that spawn terrorism by enlisting the international community to focus its efforts and resources on areas most at risk; and
- using effective public diplomacy to promote the free flow of information and ideas to kindle the hopes and aspirations of freedom of those in societies ruled by the sponsors of global terrorism.

While we recognize that our best defense is a good offense, we are also strengthening America's homeland security to protect against and deter attack. This Administration has proposed the largest government reorganization since the Truman Administration created the National Security Council and the Department of Defense. Centered on a new Department of Homeland Security and including a new unified military command and a fundamental reordering of the FBI, our comprehensive plan to secure the homeland encompasses every level of government and the cooperation of the public and the private sector.

This strategy will turn adversity into opportunity. For example, emergency management systems will be better able to cope not just with terrorism but with all hazards. Our medical system will be strengthened to manage not just bioterror, but all infectious diseases and mass-casualty dangers. Our border controls will not just stop terrorists, but improve the efficient movement of legitimate traffic.

While our focus is protecting America, we know that to defeat terrorism in today's globalized world we need support from our allies and friends. Wherever possible, the United States will rely on regional organizations and state powers to meet their obligations to fight terrorism. Where governments find the fight against terrorism beyond their capacities, we will match their willpower and their resources with whatever help we and our allies can provide.

As we pursue the terrorists in Afghanistan, we will continue to work with international organizations such as the United Nations, as well as non-governmental organizations, and other countries to provide the humanitarian, political, economic, and security assistance necessary to rebuild Afghanistan so that it will never again abuse its people, threaten its neighbors, and provide a haven for terrorists.

In the war against global terrorism, we will never forget that we are ultimately fighting for our democratic values and way of life. Freedom and fear are at war, and there will be no quick or easy end to this conflict. In leading the campaign against terrorism, we are forging new, productive international relationships and redefining existing ones in ways that meet the challenges of the twenty-first century.

IV. Work with Others to Defuse Regional Conflicts

"We build a world of justice, or we will live in a world of coercion. The magnitude of our shared responsibilities makes our disagreements look so small."

President Bush
Berlin, Germany
May 23, 2002

Concerned nations must remain actively engaged in critical regional disputes to avoid explosive escalation and minimize human suffering. In an increasingly interconnected world, regional crisis can strain our alliances, rekindle rivalries among the major powers, and create horrifying affronts to human dignity. When violence erupts and states falter, the United States will work with friends and partners to alleviate suffering and restore stability.

No doctrine can anticipate every circumstance in which U.S. action—direct or indirect—is warranted. We have finite political, economic, and military resources to meet our global priorities. The United States will approach each case with these strategic principles in mind:

- The United States should invest time and resources into building international relationships and institutions that can help manage local crises when they emerge.
- The United States should be realistic about its ability to help those who are unwilling or unready to help themselves. Where and when

people are ready to do their part, we will be willing to move decisively.

The Israeli-Palestinian conflict is critical because of the toll of human suffering, because of America's close relationship with the state of Israel and key Arab states, and because of that region's importance to other global priorities of the United States. There can be no peace for either side without freedom for both sides. America stands committed to an independent and democratic Palestine, living beside Israel in peace and security. Like all other people, Palestinians deserve a government that serves their interests and listens to their voices. The United States will continue to encourage all parties to step up to their responsibilities as we seek a just and comprehensive settlement to the conflict.

The United States, the international donor community, and the World Bank stand ready to work with a reformed Palestinian government on economic development, increased humanitarian assistance, and a program to establish, finance, and monitor a truly independent judiciary. If Palestinians embrace democracy, and the rule of law, confront corruption, and firmly reject terror, they can count on American support for the creation of a Palestinian state.

Israel also has a large stake in the success of a democratic Palestine. Permanent occupation threatens Israel's identity and democracy. So the United States continues to challenge Israeli leaders to take concrete steps to support the emergence of a viable, credible Palestinian state. As there is progress towards security, Israel forces need to withdraw fully to positions they held prior to September 28, 2000. And consistent with the recommendations of the Mitchell Committee, Israeli settlement activity in the occupied territories must stop. As violence subsides, freedom of movement should be restored, permitting innocent Palestinians to resume work and normal life. The United States can play a crucial role but, ultimately, lasting peace can only come when Israelis and Palestinians resolve the issues and end the conflict between them.

In South Asia, the United States has also emphasized the need for India and Pakistan to resolve their disputes. This Administration invested time and resources building strong bilateral relations with India and Pakistan. These strong relations then gave us leverage to play a constructive role when tensions in the region became acute. With Pakistan, our bilateral relations have been bolstered by Pakistan's choice to join the war against terror and move toward building a more open and tolerant society. The Administration sees India's potential to become one of the great democratic powers of the twenty-first century and has worked hard to transform our relationship accordingly. Our involvement in this regional dispute, building on earlier investments in bilateral relations, looks first to concrete steps by India and Pakistan that can help defuse military confrontation.

Indonesia took courageous steps to create a working democracy and respect for the rule of law. By tolerating ethnic minorities, respecting the rule of law, and accepting open markets, Indonesia may be able to employ the engine of opportunity that has helped lift some of its neighbors out of poverty and desperation. It is the initiative by Indonesia that allows U.S. assistance to make a difference.

In the Western Hemisphere we have formed flexible coalitions with countries that share our priorities, particularly Mexico, Brazil, Canada, Chile, and Colombia. Together we will promote a truly democratic hemisphere where our integration advances security, prosperity, opportunity, and hope. We will work with regional institutions, such as the Summit of the Americas process, the Organization of American States (OAS), and the Defense Ministerial of the Americas for the benefit of the entire hemisphere.

Parts of Latin America confront regional conflict, especially arising from the violence of drug cartels and their accomplices. This conflict and unrestrained narcotics trafficking could imperil the health and security of the United States. Therefore we have developed an active strategy to help the Andean nations adjust their economies, enforce their laws, defeat terrorist organizations, and cut off the supply of drugs, while-as important-we work to reduce the demand for drugs in our own country.

In Colombia, we recognize the link between terrorist and extremist groups that challenge the security of the state and drug trafficking activities that help finance the operations of such groups. We are working to help Colombia defend its democratic institutions and defeat illegal armed groups of both the left and right by extending effective sovereignty over the entire national territory and provide basic security to the Colombian people.

In Africa, promise and opportunity sit side by side with disease, war, and desperate poverty. This threatens both a core value of the United States—preserving human dignity—and our strategic priority—combating global terror. American interests and American principles, therefore, lead in the same direction: we will work with others for an African continent that lives in liberty, peace, and growing prosperity. Together with our European allies, we must help strengthen Africa's fragile states, help build indigenous capability to secure porous borders, and help build up the law enforcement and intelligence infrastructure to deny havens for terrorists.

An ever more lethal environment exists in Africa as local civil wars spread beyond borders to create regional war zones. Forming coalitions of the willing and cooperative security arrangements are key to confronting these emerging transnational threats.

Africa's great size and diversity requires a security strategy that focuses on bilateral engagement and builds coalitions of the willing. This Administration will focus on three interlocking strategies for the region:

- countries with major impact on their neighborhood such as South Africa, Nigeria, Kenya, and Ethiopia are anchors for regional engagement and require focused attention;
- coordination with European allies and international institutions is essential for constructive conflict mediation and successful peace operations; and
- Africa's capable reforming states and sub-regional organizations must be strengthened as the primary means to address transnational threats on a sustained basis.

Ultimately the path of political and economic freedom presents the surest route to progress in sub-Saharan Africa, where most wars are conflicts over material resources and political access often tragically waged on the basis of ethnic and religious difference. The transition to the African Union with its stated commitment to good governance and a common responsibility for democratic political systems offers opportunities to strengthen democracy on the continent.

V. Prevent Our Enemies from Threatening Us, Our Allies, and Our Friends with Weapons of Mass Destruction

"The gravest danger to freedom lies at the crossroads of radicalism and technology. When the spread of chemical and biological and nuclear weapons, along with ballistic missile technology—when that occurs, even weak states and small groups could attain a catastrophic power to strike great nations. Our enemies have declared this very intention, and have been caught seeking these terrible weapons. They want the capability to blackmail us, or to harm us, or to harm our friends—and we will oppose them with all our power."

President Bush
West Point, New York
June 1, 2002

The nature of the Cold War threat required the United States—with our allies and friends—to emphasize deterrence of the enemy's use of force, producing a grim strategy of mutual assured destruction. With the collapse of the Soviet Union and the end of the Cold War, our security environment has undergone profound transformation.

Having moved from confrontation to cooperation as the hallmark of our relationship with Russia, the dividends are evident: an end to the balance of terror that divided us; an historic reduction in the nuclear arsenals on both sides; and cooperation in areas such as counterterrorism and missile defense that until recently were inconceivable.

But new deadly challenges have emerged from rogue states and terrorists. None of these contemporary threats rival the sheer destructive power that was arrayed against us by the Soviet Union. However, the nature and motivations of these new adversaries, their determination to obtain destructive powers hitherto available only to the world's strongest states, and the greater likelihood that they will use weapons of mass destruction against us, make today's security environment more complex and dangerous.

In the 1990s we witnessed the emergence of a small number of rogue states that, while different in important ways, share a number of attributes. These states:

- brutalize their own people and squander their national resources for the personal gain of the rulers;
- display no regard for international law, threaten their neighbors, and callously violate international treaties to which they are party;
- are determined to acquire weapons of mass destruction, along with other advanced military technology, to be used as threats or offensively to achieve the aggressive designs of these regimes;
- sponsor terrorism around the globe; and
- reject basic human values and hate the United States and everything for which it stands.

At the time of the Gulf War, we acquired irrefutable proof that Iraq's designs were not limited to the chemical weapons it had used against Iran and its own people, but also extended to the acquisition of nuclear weapons and biological agents. In the past decade North Korea has become the world's principal purveyor of ballistic missiles, and has tested increasingly capable missiles while developing its own WMD arsenal. Other rogue regimes seek nuclear, biological, and chemical weapons as well. These states' pursuit of, and global trade in, such weapons has become a looming threat to all nations.

We must be prepared to stop rogue states and their terrorist clients before they are able to threaten or use weapons of mass destruction against the United States and our allies and friends. Our response must take full advantage of strengthened alliances, the establishment of new partnerships with former adversaries, innovation in the use of military forces, modern technologies, including the development of an effective missile defense system, and increased emphasis on intelligence collection and analysis.

Our comprehensive strategy to combat WMD includes:

- *Proactive counterproliferation efforts.* We must deter and defend against the threat before it is unleashed. We must ensure that key capabilities—detection, active and passive defenses, and counter-force capabilities—are integrated into our defense transformation and our homeland security systems. Counterproliferation must also be integrated into the doctrine, training, and equipping of our forces and those of our allies to ensure that we can prevail in any conflict with WMD-armed adversaries.

- *Strengthened nonproliferation efforts to prevent rogue states and terrorists from acquiring the materials, technologies, and expertise necessary for weapons of mass destruction.* We will enhance diplomacy, arms control, multilateral export controls, and threat reduction assistance that impede states and terrorists seeking WMD, and when necessary, interdict enabling technologies and materials. We will continue to build coalitions to support these efforts, encouraging their increased political and financial support for nonproliferation and threat reduction programs. The recent G-8 agreement to commit up to $20 billion to a global partnership against proliferation marks a major step forward.

- *Effective consequence management to respond to the effects of WMD use, whether by terrorists or hostile states.* Minimizing the effects of WMD use against our people will help deter those who possess such weapons and dissuade those who seek to acquire them by persuading enemies that they cannot attain their desired ends. The United States must also be prepared to respond to the effects of WMD use against our forces abroad, and to help friends and allies if they are attacked.

It has taken almost a decade for us to comprehend the true nature of this new threat. Given the goals of rogue states and terrorists, the United States can no longer solely rely on a reactive posture as we have in the past. The inability to deter a potential attacker, the immediacy of today's threats, and the magnitude of potential harm that could be caused by our adversaries' choice of weapons, do not permit that option. We cannot let our enemies strike first.

In the Cold War, especially following the Cuban missile crisis, we faced a generally status quo, risk-averse adversary. Deterrence was an effective defense. But deterrence based only upon the threat of retaliation is less likely to work against leaders of rogue states more willing to take risks, gambling with the lives of their people, and the wealth of their nations.

- In the Cold War, weapons of mass destruction were considered weapons of last resort whose use risked the destruction of those who used them. Today, our enemies see weapons of mass destruction as weapons of choice. For rogue states these weapons are tools of intimidation and military aggression against their neighbors. These weapons may also allow these states to attempt to blackmail the United States and our allies to prevent us from deterring or repelling the aggressive behavior of rogue states. Such states also see these weapons as their best means of overcoming the conventional superiority of the United States.
- Traditional concepts of deterrence will not work against a terrorist enemy whose avowed tactics are wanton destruction and the targeting of innocents; whose so-called soldiers seek martyrdom in death and whose most potent protection is statelessness. The overlap between states that sponsor terror and those that pursue WMD compels us to action.

For centuries, international law recognized that nations need not suffer an attack before they can lawfully take action to defend themselves against forces that present an imminent danger of attack. Legal scholars and international jurists often conditioned the legitimacy of preemption on the existence of an imminent threat—most often a visible mobilization of armies, navies, and air forces preparing to attack.

We must adapt the concept of imminent threat to the capabilities and objectives of today's adversaries. Rogue states and terrorists do not seek to attack us using conventional means. They know such attacks would fail. Instead, they rely on acts of terror and, potentially, the use of weapons of mass destruction—weapons that can be easily concealed, delivered covertly, and used without warning.

The targets of these attacks are our military forces and our civilian population, in direct violation of one of the principal norms of the law of warfare. As was demonstrated by the losses on September 11, 2001, mass civilian casualties is the specific objective of terrorists and these losses would be exponentially more severe if terrorists acquired and used weapons of mass destruction.

The United States has long maintained the option of preemptive actions to counter a sufficient threat to our national security. The greater the threat, the greater is the risk of inaction——and the more compelling the case for taking anticipatory action to defend ourselves, even if uncertainty remains as to the time and place of the enemy's attack. To forestall or prevent such hostile acts by our adversaries, the United States will, if necessary, act preemptively.

The United States will not use force in all cases to preempt emerging threats, nor should nations use preemption as a pretext for aggression. Yet in an age where the enemies of civilization openly and actively seek the world's most destructive technologies, the United States cannot remain idle while dangers gather. We will always proceed deliberately, weighing the consequences of our actions. To support preemptive options, we will:

- build better, more integrated intelligence capabilities to provide timely, accurate information on threats, wherever they may emerge;
- coordinate closely with allies to form a common assessment of the most dangerous threats; and
- continue to transform our military forces to ensure our ability to conduct rapid and precise operations to achieve decisive results.

The purpose of our actions will always be to eliminate a specific threat to the United States or our allies and friends. The reasons for our actions will be clear, the force measured, and the cause just.

VI. Ignite a New Era of Global Economic Growth through Free Markets and Free Trade

"When nations close their markets and opportunity is hoarded by a privileged few, no amount—no amount—of development aid is ever enough. When nations respect their people, open markets, invest in better health and education, every dollar of aid, every dollar of trade revenue and domestic capital is used more effectively."

President Bush
Monterrey, Mexico
March 22, 2002

A strong world economy enhances our national security by advancing prosperity and freedom in the rest of the world. Economic growth supported by free trade and free markets creates new jobs and higher incomes. It allows people to lift their lives out of poverty, spurs economic and legal reform, and the fight against corruption, and it reinforces the habits of liberty.

We will promote economic growth and economic freedom beyond America's shores. All governments are responsible for creating their own economic policies and responding to their own economic challenges. We will use our economic engagement with other countries to underscore the benefits of policies that generate higher productivity and sustained economic growth, including:

- pro-growth legal and regulatory policies to encourage business investment, innovation, and entrepreneurial activity;
- tax policies—particularly lower marginal tax rates—that improve incentives for work and investment;
- rule of law and intolerance of corruption so that people are confident that they will be able to enjoy the fruits of their economic endeavors;
- strong financial systems that allow capital to be put to its most efficient use;
- sound fiscal policies to support business activity;
- investments in health and education that improve the well-being and skills of the labor force and population as a whole; and
- free trade that provides new avenues for growth and fosters the diffusion of technologies and ideas that increase productivity and opportunity.

The lessons of history are clear: market economies, not command-and-control economies with the heavy hand of government, are the best way to promote prosperity and reduce poverty. Policies that further strengthen market incentives and market institutions are relevant for all economies—industrialized countries, emerging markets, and the developing world.

A return to strong economic growth in Europe and Japan is vital to U.S. national security interests. We want our allies to have strong economies for their own sake, for the sake of the global economy, and for the sake of global security. European efforts to remove structural barriers in their economies are particularly important in this regard, as are Japan's efforts to end deflation and address the problems of non-performing loans in the Japanese banking system. We will continue to use our regular consultations with Japan and our European partners—including through the Group of Seven (G-7)—to discuss policies they are adopting to promote growth in their economies and support higher global economic growth.

Improving stability in emerging markets is also key to global economic growth. International flows of investment capital are needed to expand the productive potential of these economies. These flows allow emerging markets and developing countries to make the investments that raise living standards and reduce poverty. Our long-term objective should be a world in which all countries have investment-grade credit ratings that allow them access to international capital markets and to invest in their future.

We are committed to policies that will help emerging markets achieve access to larger capital flows at lower cost. To this end, we will continue to pursue reforms aimed at reducing uncertainty in financial markets. We will work actively with other countries, the International Monetary Fund (IMF), and the private sector to implement the G-7 Action Plan negotiated earlier this year for preventing financial crises and more effectively resolving them when they occur.

The best way to deal with financial crises is to prevent them from occurring, and we have encouraged the IMF to improve its efforts doing so. We will continue to work with the IMF to streamline the policy conditions for its lending and to focus its lending strategy on achieving economic growth through sound fiscal and monetary policy, exchange rate policy, and financial sector policy.

The concept of "free trade" arose as a moral principle even before it became a pillar of economics. If you can make something that others value, you should be able to sell it to them. If others make something that you value, you should be able to buy it. This is real freedom, the freedom for a person—or a nation—to make a living. To promote free trade, the Unites States has developed a comprehensive strategy:

- *Seize the global initiative.* The new global trade negotiations we helped launch at Doha in November 2001 will have an ambitious agenda, especially in agriculture, manufacturing, and services, targeted for completion in 2005. The United States has led the way in completing the accession of China and a democratic Taiwan to the World Trade Organization. We will assist Russia's preparations to join the WTO.
- *Press regional initiatives.* The United States and other democracies in the Western Hemisphere have agreed to create the Free Trade Area of the Americas, targeted for completion in 2005. This year the United States will advocate market-access negotiations with its partners,

targeted on agriculture, industrial goods, services, investment, and government procurement. We will also offer more opportunity to the poorest continent, Africa, starting with full use of the preferences allowed in the African Growth and Opportunity Act, and leading to free trade.

- *Move ahead with bilateral free trade agreements.* Building on the free trade agreement with Jordan enacted in 2001, the Administration will work this year to complete free trade agreements with Chile and Singapore. Our aim is to achieve free trade agreements with a mix of developed and developing countries in all regions of the world. Initially, Central America, Southern Africa, Morocco, and Australia will be our principal focal points.

- *Renew the executive-congressional partnership.* Every administration's trade strategy depends on a productive partnership with Congress. After a gap of 8 years, the Administration reestablished majority support in the Congress for trade liberalization by passing Trade Promotion Authority and the other market opening measures for developing countries in the Trade Act of 2002. This Administration will work with Congress to enact new bilateral, regional, and global trade agreements that will be concluded under the recently passed Trade Promotion Authority.

- *Promote the connection between trade and development.* Trade policies can help developing countries strengthen property rights, competition, the rule of law, investment, the spread of knowledge, open societies, the efficient allocation of resources, and regional integration—all leading to growth, opportunity, and confidence in developing countries. The United States is implementing The Africa Growth and Opportunity Act to provide market-access for nearly all goods produced in the 35 countries of sub-Saharan Africa. We will make more use of this act and its equivalent for the Caribbean Basin and continue to work with multilateral and regional institutions to help poorer countries take advantage of these opportunities. Beyond market access, the most important area where trade intersects with poverty is in public health. We will ensure that the WTO intellectual property rules are flexible enough to allow developing nations to gain access to critical medicines for extraordinary dangers like HIV/AIDS, tuberculosis, and malaria.

- *Enforce trade agreements and laws against unfair practices. Commerce depends on the rule of law; international trade depends on enforceable agreements.* Our top priorities are to resolve ongoing disputes with the European Union, Canada, and Mexico and to make a global effort to address new technology, science, and health regulations that needlessly impede farm exports and improved agriculture. Laws against unfair trade practices are often abused, but the international community must be able to address genuine concerns about government subsidies and dumping. International industrial espionage which undermines fair competition must be detected and deterred.

- *Help domestic industries and workers adjust.* There is a sound statutory framework for these transitional safeguards which we have used in the agricultural sector and which we are using this year to help the American steel industry. The benefits of free trade depend upon the enforcement of fair trading practices. These safeguards help ensure that the benefits of free trade do not come at the expense of American workers. Trade adjustment assistance will help workers adapt to the change and dynamism of open markets.

- *Protect the environment and workers.* The United States must foster economic growth in ways that will provide a better life along with widening prosperity. We will incorporate labor and environmental concerns into U.S. trade negotiations, creating a healthy "network" between multilateral environmental agreements with the WTO, and use the International Labor Organization, trade preference programs, and trade talks to improve working conditions in conjunction with freer trade.

- *Enhance energy security.* We will strengthen our own energy security and the shared prosperity of the global economy by working with our allies, trading partners, and energy producers to expand the sources and types of global energy supplied, especially in the Western Hemisphere, Africa, Central Asia, and the Caspian region. We will also continue to work with our partners to develop cleaner and more energy efficient technologies.

Economic growth should be accompanied by global efforts to stabilize greenhouse gas concentrations associated with this growth, containing them at a level that prevents dangerous human interference with the global climate. Our overall objective is to reduce America's greenhouse gas emissions relative to the size of our economy, cutting such emissions per unit of economic activity by 18 percent over the next 10 years, by the year 2012. Our strategies for attaining this goal will be to:

- remain committed to the basic U.N. Framework Convention for international cooperation;

- obtain agreements with key industries to cut emissions of some of the most potent greenhouse gases and give transferable credits to companies that can show real cuts;
- develop improved standards for measuring and registering emission reductions;
- promote renewable energy production and clean coal technology, as well as nuclear power—which produces no greenhouse gas emissions, while also improving fuel economy for U.S. cars and trucks;
- increase spending on research and new conservation technologies, to a total of $4.5 billion—the largest sum being spent on climate change by any country in the world and a $700 million increase over last year's budget; and
- assist developing countries, especially the major greenhouse gas emitters such as China and India, so that they will have the tools and resources to join this effort and be able to grow along a cleaner and better path.

VII. Expand the Circle of Development by Opening Societies and Building the Infrastructure of Democracy

"In World War II we fought to make the world safer, then worked to rebuild it. As we wage war today to keep the world safe from terror, we must also work to make the world a better place for all its citizens."

President Bush
Washington, D.C. (Inter-American Development Bank)
March 14, 2002

A world where some live in comfort and plenty, while half of the human race lives on less than $2 a day, is neither just nor stable. Including all of the world's poor in an expanding circle of development—and opportunity— is a moral imperative and one of the top priorities of U.S. international policy.

Decades of massive development assistance have failed to spur economic growth in the poorest countries. Worse, development aid has often served to prop up failed policies, relieving the pressure for reform and perpetuating misery. Results of aid are typically measured in dollars spent by donors, not in the rates of growth and poverty reduction achieved by recipients. These are the indicators of a failed strategy.

Working with other nations, the United States is confronting this failure. We forged a new consensus at the U.N. Conference on Financing for

Development in Monterrey that the objectives of assistance—and the strategies to achieve those objectives—must change.

This Administration's goal is to help unleash the productive potential of individuals in all nations. Sustained growth and poverty reduction is impossible without the right national policies. Where governments have implemented real policy changes, we will provide significant new levels of assistance. The United States and other developed countries should set an ambitious and specific target: to double the size of the world's poorest economies within a decade.

The United States Government will pursue these major strategies to achieve this goal:

- *Provide resources to aid countries that have met the challenge of national reform.* We propose a 50 percent increase in the core development assistance given by the United States. While continuing our present programs, including humanitarian assistance based on need alone, these billions of new dollars will form a new Millennium Challenge Account for projects in countries whose governments rule justly, invest in their people, and encourage economic freedom. Governments must fight corruption, respect basic human rights, embrace the rule of law, invest in health care and education, follow responsible economic policies, and enable entrepreneurship. The Millennium Challenge Account will reward countries that have demonstrated real policy change and challenge those that have not to implement reforms.
- *Improve the effectiveness of the World Bank and other development banks in raising living standards.* The United States is committed to a comprehensive reform agenda for making the World Bank and the other multilateral development banks more effective in improving the lives of the world's poor. We have reversed the downward trend in U.S. contributions and proposed an 18 percent increase in the U.S. contributions to the International Development Association (IDA)— the World Bank's fund for the poorest countries—and the African Development Fund. The key to raising living standards and reducing poverty around the world is increasing productivity growth, especially in the poorest countries. We will continue to press the multilateral development banks to focus on activities that increase economic productivity, such as improvements in education, health, rule of law, and private sector development. Every project, every loan, every grant must be judged by how much it will increase productivity growth in developing countries.

- *Insist upon measurable results to ensure that development assistance is actually making a difference in the lives of the world's poor.* When it comes to economic development, what really matters is that more children are getting a better education, more people have access to health care and clean water, or more workers can find jobs to make a better future for their families. We have a moral obligation to measure the success of our development assistance by whether it is delivering results. For this reason, we will continue to demand that our own development assistance as well as assistance from the multilateral development banks has measurable goals and concrete benchmarks for achieving those goals. Thanks to U.S. leadership, the recent IDA replenishment agreement will establish a monitoring and evaluation system that measures recipient countries' progress. For the first time, donors can link a portion of their contributions to IDA to the achievement of actual development results, and part of the U.S. contribution is linked in this way. We will strive to make sure that the World Bank and other multilateral development banks build on this progress so that a focus on results is an integral part of everything that these institutions do.

- *Increase the amount of development assistance that is provided in the form of grants instead of loans.* Greater use of results-based grants is the best way to help poor countries make productive investments, particularly in the social sectors, without saddling them with ever-larger debt burdens. As a result of U.S. leadership, the recent IDA agreement provided for significant increases in grant funding for the poorest countries for education, HIV/AIDS, health, nutrition, water, sanitation, and other human needs. Our goal is to build on that progress by increasing the use of grants at the other multilateral development banks. We will also challenge universities, nonprofits, and the private sector to match government efforts by using grants to support development projects that show results.

- *Open societies to commerce and investment. Trade and investment are the real engines of economic growth.* Even if government aid increases, most money for development must come from trade, domestic capital, and foreign investment. An effective strategy must try to expand these flows as well. Free markets and free trade are key priorities of our national security strategy.

- *Secure public health.* The scale of the public health crisis in poor countries is enormous. In countries afflicted by epidemics and pandemics like HIV/AIDS, malaria, and tuberculosis, growth and development will be threatened until these scourges can be contained.

Resources from the developed world are necessary but will be effective only with honest governance, which supports prevention programs and provides effective local infrastructure. The United States has strongly backed the new global fund for HIV/AIDS organized by U.N. Secretary General Kofi Annan and its focus on combining prevention with a broad strategy for treatment and care. The United States already contributes more than twice as much money to such efforts as the next largest donor. If the global fund demonstrates its promise, we will be ready to give even more.

- *Emphasize education.* Literacy and learning are the foundation of democracy and development. Only about 7 percent of World Bank resources are devoted to education. This proportion should grow. The United States will increase its own funding for education assistance by at least 20 percent with an emphasis on improving basic education and teacher training in Africa. The United States can also bring information technology to these societies, many of whose education systems have been devastated by HIV/AIDS.

- *Continue to aid agricultural development.* New technologies, including biotechnology, have enormous potential to improve crop yields in developing countries while using fewer pesticides and less water. Using sound science, the United States should help bring these benefits to the 800 million people, including 300 million children, who still suffer from hunger and malnutrition.

VIII. Develop Agendas for Cooperative Action with the Other Main Centers of Global Power

"We have our best chance since the rise of the nation-state in the 17th century to build a world where the great powers compete in peace instead of prepare for war."

President Bush
West Point, New York
June 1, 2002

America will implement its strategies by organizing coalitions—as broad as practicable—of states able and willing to promote a balance of power that favors freedom. Effective coalition leadership requires clear priorities, an appreciation of others' interests, and consistent consultations among partners with a spirit of humility.

There is little of lasting consequence that the United States can accomplish in the world without the sustained cooperation of its allies and friends in Canada and Europe. Europe is also the seat of two of the strongest and most able international institutions in the world: the North Atlantic Treaty Organization (NATO), which has, since its inception, been the fulcrum of transatlantic and inter-European security, and the European Union (EU), our partner in opening world trade.

The attacks of September 11 were also an attack on NATO, as NATO itself recognized when it invoked its Article V self-defense clause for the first time. NATO's core mission—collective defense of the transatlantic alliance of democracies—remains, but NATO must develop new structures and capabilities to carry out that mission under new circumstances. NATO must build a capability to field, at short notice, highly mobile, specially trained forces whenever they are needed to respond to a threat against any member of the alliance.

The alliance must be able to act wherever our interests are threatened, creating coalitions under NATO's own mandate, as well as contributing to mission-based coalitions. To achieve this, we must:

- expand NATO's membership to those democratic nations willing and able to share the burden of defending and advancing our common interests;
- ensure that the military forces of NATO nations have appropriate combat contributions to make in coalition warfare;
- develop planning processes to enable those contributions to become effective multinational fighting forces;
- take advantage of the technological opportunities and economies of scale in our defense spending to transform NATO military forces so that they dominate potential aggressors and diminish our vulnerabilities;
- streamline and increase the flexibility of command structures to meet new operational demands and the associated requirements of training, integrating, and experimenting with new force configurations; and
- maintain the ability to work and fight together as allies even as we take the necessary steps to transform and modernize our forces.

If NATO succeeds in enacting these changes, the rewards will be a partnership as central to the security and interests of its member states as was the case during the Cold War. We will sustain a common perspective on the threats to our societies and improve our ability to take common action in defense of our nations and their interests. At the same time, we welcome our European allies' efforts to forge a greater foreign policy and defense identity with the EU, and commit ourselves to close consultations to ensure that these developments work with NATO. We cannot afford to lose this opportunity to better prepare the family of transatlantic democracies for the challenges to come.

The attacks of September 11 energized America's Asian alliances. Australia invoked the ANZUS Treaty to declare the September 11 was an attack on Australia itself, following that historic decision with the dispatch of some of the world's finest combat forces for Operation Enduring Freedom. Japan and the Republic of Korea provided unprecedented levels of military logistical support within weeks of the terrorist attack. We have deepened cooperation on counterterrorism with our alliance partners in Thailand and the Philippines and received invaluable assistance from close friends like Singapore and New Zealand.

The war against terrorism has proven that America's alliances in Asia not only underpin regional peace and stability, but are flexible and ready to deal with new challenges. To enhance our Asian alliances and friendships, we will:

- look to Japan to continue forging a leading role in regional and global affairs based on our common interests, our common values, and our close defense and diplomatic cooperation;
- work with South Korea to maintain vigilance towards the North while preparing our alliance to make contributions to the broader stability of the region over the longer term;
- build on 50 years of U.S.-Australian alliance cooperation as we continue working together to resolve regional and global problems—as we have so many times from the Battle of the Coral Sea to Tora Bora;
- maintain forces in the region that reflect our commitments to our allies, our requirements, our technological advances, and the strategic environment; and
- build on stability provided by these alliances, as well as with institutions such as ASEAN and the Asia-Pacific Economic Cooperation forum, to develop a mix of regional and bilateral strategies to manage change in this dynamic region.

We are attentive to the possible renewal of old patterns of great power competition. Several potential great powers are now in the midst of internal transition-most importantly Russia, India, and China. In all three cases, recent developments have encouraged our hope that a truly global consensus about basic principles is slowly taking shape.

With Russia, we are already building a new strategic relationship based on a central reality of the twenty-first century: the United States and Russia are no longer strategic adversaries. The Moscow Treaty on Strategic Reductions is emblematic of this new reality and reflects a critical change in Russian thinking that promises to lead to productive, long-term relations with the Euro-Atlantic community and the United States. Russia's top leaders have a realistic assessment of their country's current weakness and the policies—internal and external—needed to reverse those weaknesses. They understand, increasingly, that Cold War approaches do not serve their national interests and that Russian and American strategic interests overlap in many areas.

United States policy seeks to use this turn in Russian thinking to refocus our relationship on emerging and potential common interests and challenges. We are broadening our already extensive cooperation in the global war on terrorism. We are facilitating Russia's entry into the World Trade Organization, without lowering standards for accession, to promote beneficial bilateral trade and investment relations. We have created the NATO-Russia Council with the goal of deepening security cooperation among Russia, our European allies, and ourselves. We will continue to bolster the independence and stability of the states of the former Soviet Union in the belief that a prosperous and stable neighborhood will reinforce Russia's growing commitment to integration into the Euro-Atlantic community.

At the same time, we are realistic about the differences that still divide us from Russia and about the time and effort it will take to build an enduring strategic partnership. Lingering distrust of our motives and policies by key Russian elites slows improvement in our relations. Russia's uneven commitment to the basic values of free-market democracy and dubious record in combating the proliferation of weapons of mass destruction remain matters of great concern. Russia's very weakness limits the opportunities for cooperation. Nevertheless, those opportunities are vastly greater now than in recent years—or even decades.

The United States has undertaken a transformation in its bilateral relationship with India based on a conviction that U.S. interests require a strong relationship with India. We are the two largest democracies, committed to political freedom protected by representative government. India is moving toward greater economic freedom as well. We have a common interest in the free flow of commerce, including through the vital sea lanes of the Indian Ocean. Finally, we share an interest in fighting terrorism and in creating a strategically stable Asia.

Differences remain, including over the development of India's nuclear and missile programs, and the pace of India's economic reforms. But while in the past these concerns may have dominated our thinking about India, today we start with a view of India as a growing world power with which we have common strategic interests. Through a strong partnership with India, we can best address any differences and shape a dynamic future.

The United States relationship with China is an important part of our strategy to promote a stable, peaceful, and prosperous Asia-Pacific region. We welcome the emergence of a strong, peaceful, and prosperous China. The democratic development of China is crucial to that future. Yet, a quarter century after beginning the process of shedding the worst features of the Communist legacy, China's leaders have not yet made the next series of fundamental choices about the character of their state. In pursuing advanced military capabilities that can threaten its neighbors in the Asia-Pacific region, China is following an outdated path that, in the end, will hamper its own pursuit of national greatness. In time, China will find that social and political freedom is the only source of that greatness.

The United States seeks a constructive relationship with a changing China. We already cooperate well where our interests overlap, including the current war on terrorism and in promoting stability on the Korean peninsula. Likewise, we have coordinated on the future of Afghanistan and have initiated a comprehensive dialogue on counterterrorism and similar transitional concerns. Shared health and environmental threats, such as the spread of HIV/AIDS, challenge us to promote jointly the welfare of our citizens.

Addressing these transnational threats will challenge China to become more open with information, promote the development of civil society, and enhance individual human rights. China has begun to take the road to political openness, permitting many personal freedoms and conducting village-level elections, yet remains strongly committed to national one-party rule by the Communist Party. To make that nation truly accountable to its citizen's needs and aspirations, however, much work remains to be done. Only by allowing the Chinese people to think, assemble, and worship freely can China reach its full potential.

Our important trade relationship will benefit from China's entry into the World Trade Organization, which will create more export opportunities and ultimately more jobs for American farmers, workers, and companies. China is our fourth largest trading partner, with over $100 billion in annual two-way trade. The power of market principles and the WTO's requirements for transparency and accountability will advance openness and the rule of law in China to help establish basic protections for commerce and for citizens. There are, however, other areas in which we have profound disagreements. Our commitment to the self-defense of Taiwan under the Taiwan Relations Act is one. Human rights is another. We expect China to adhere to its nonproliferation commitments. We will work to narrow differences where they exist, but not allow them to preclude cooperation where we agree.

The events of September 11, 2001, fundamentally changed the context for relations between the United States and other main centers of global power, and opened vast, new opportunities. With our long-standing allies in Europe and Asia, and with leaders in Russia, India, and China, we must develop active agendas of cooperation lest these relationships become routine and unproductive.

Every agency of the United States Government shares the challenge. We can build fruitful habits of consultation, quiet argument, sober analysis, and common action. In the long-term, these are the practices that will sustain the supremacy of our common principles and keep open the path of progress.

IX. Transform America's National Security Institutions to Meet the Challenges and Opportunities of the Twenty-First Century

> *"Terrorists attacked a symbol of American prosperity. They did not touch its source. America is successful because of the hard work, creativity, and enterprise of our people."*
>
> President Bush
> Washington, D.C. (Joint Session of Congress)
> September 20, 2001

The major institutions of American national security were designed in a different era to meet different requirements. All of them must be transformed.

It is time to reaffirm the essential role of American military strength. We must build and maintain our defenses beyond challenge. Our military's highest priority is to defend the United States. To do so effectively, our military must:

- assure our allies and friends;
- dissuade future military competition;
- deter threats against U.S. interests, allies, and friends; and
- decisively defeat any adversary if deterrence fails.

The unparalleled strength of the United States armed forces, and their forward presence, have maintained the peace in some of the world's most strategically vital regions. However, the threats and enemies we must confront have changed, and so must our forces. A military structured to deter massive Cold War-era armies must be transformed to focus more on how an adversary might fight rather than where and when a war might occur. We will channel our energies to overcome a host of operational challenges.

The presence of American forces overseas is one of the most profound symbols of the U.S. commitments to allies and friends. Through our willingness to use force in our own defense and in defense of others, the United States demonstrates its resolve to maintain a balance of power that favors freedom. To contend with uncertainty and to meet the many security challenges we face, the United States will require bases and stations within and beyond Western Europe and Northeast Asia, as well as temporary access arrangements for the long-distance deployment of U.S. forces.

Before the war in Afghanistan, that area was low on the list of major planning contingencies. Yet, in a very short time, we had to operate across the length and breadth of that remote nation, using every branch of the armed forces. We must prepare for more such deployments by developing assets such as advanced remote sensing, long-range precision strike capabilities, and transformed maneuver and expeditionary forces. This broad portfolio of military capabilities must also include the ability to defend the homeland, conduct information operations, ensure U.S. access to distant theaters, and protect critical U.S. infrastructure and assets in outer space.

Innovation within the armed forces will rest on experimentation with new approaches to warfare, strengthening joint operations, exploiting U.S. intelligence advantages, and taking full advantage of science and technology. We must also transform the way the Department of Defense is run, especially in financial management and recruitment and retention. Finally, while maintaining near-term readiness and the ability to fight the war on terrorism, the goal must be to provide the President with a wider range of military options to discourage aggression or any form of coercion against the United States, our allies, and our friends.

We know from history that deterrence can fail; and we know from experience that some enemies cannot be deterred. The United States must and will maintain the capability to defeat any attempt by an enemy—whether a state or non-state actor—to impose its will on the United States, our allies, or our friends. We will maintain the forces sufficient to support our obligations, and to defend freedom. Our forces will be strong enough to dissuade potential adversaries from pursuing a military build-up in hopes of surpassing, or equaling, the power of the United States.

Intelligence—and how we use it—is our first line of defense against terrorists and the threat posed by hostile states. Designed around the priority of gathering enormous information about a massive, fixed object-the Soviet bloc—the intelligence community is coping with the challenge of following a far more complex and elusive set of targets.

We must transform our intelligence capabilities and build new ones to keep pace with the nature of these threats. Intelligence must be appropriately integrated with our defense and law enforcement systems and coordinated with our allies and friends. We need to protect the capabilities we have so that we do not arm our enemies with the knowledge of how best to surprise us. Those who would harm us also seek the benefit of surprise to limit our prevention and response options and to maximize injury.

We must strengthen intelligence warning and analysis to provide integrated threat assessments for national and homeland security. Since the threats inspired by foreign governments and groups may be conducted inside the United States, we must also ensure the proper fusion of information between intelligence and law enforcement.

Initiatives in this area will include:

- strengthening the authority of the Director of Central Intelligence to lead the development and actions of the Nation's foreign intelligence capabilities;
- establishing a new framework for intelligence warning that provides seamless and integrated warning across the spectrum of threats facing the nation and our allies;
- continuing to develop new methods of collecting information to sustain our intelligence advantage;
- investing in future capabilities while working to protect them through a more vigorous effort to prevent the compromise of intelligence capabilities; and
- collecting intelligence against the terrorist danger across the government with allsource analysis.

As the United States Government relies on the armed forces to defend America's interests, it must rely on diplomacy to interact with other nations. We will ensure that the Department of State receives funding sufficient to ensure the success of American diplomacy. The State Department takes the lead in managing our bilateral relationships with other governments. And in this new era, its people and institutions must be able to interact equally adroitly with non-governmental organizations and international institutions. Officials trained mainly in international politics must also extend their reach to understand complex issues of domestic governance around the world, including public health, education, law enforcement, the judiciary, and public diplomacy.

Our diplomats serve at the front line of complex negotiations, civil wars, and other humanitarian catastrophes. As humanitarian relief requirements are better understood, we must also be able to help build police forces, court systems, and legal codes, local and provincial government institutions, and electoral systems. Effective international cooperation is needed to accomplish these goals, backed by American readiness to play our part.

Just as our diplomatic institutions must adapt so that we can reach out to others, we also need a different and more comprehensive approach to public information efforts that can help people around the world learn about and understand America. The war on terrorism is not a clash of civilizations. It does, however, reveal the clash inside a civilization, a battle for the future of the Muslim world. This is a struggle of ideas and this is an area where America must excel.

We will take the actions necessary to ensure that our efforts to meet our global security commitments and protect Americans are not impaired by the potential for investigations, inquiry, or prosecution by the International Criminal Court (ICC), whose jurisdiction does not extend to Americans and which we do not accept. We will work together with other nations to avoid complications in our military operations and cooperation, through such mechanisms as multilateral and bilateral agreements that will protect U.S. nationals from the ICC. We will implement fully the American Servicemembers Protection Act, whose provisions are intended to ensure and enhance the protection of U.S. personnel and officials.

We will make hard choices in the coming year and beyond to ensure the right level and allocation of government spending on national security. The United States Government must strengthen its defenses to win this war. At home, our most important priority is to protect the homeland for the American people.

Today, the distinction between domestic and foreign affairs is diminishing. In a globalized world, events beyond America's borders have a greater impact inside them. Our society must be open to people, ideas, and goods from across the globe. The characteristics we most cherish—our freedom, our cities, our systems of movement, and modern life—are vulnerable to terrorism. This vulnerability will persist long after we bring to justice those responsible for the September 11 attacks. As time passes, individuals may gain access to means of destruction that until now could be wielded only by armies, fleets, and squadrons. This is a new condition of life. We will adjust to it and thrive—in spite of it.

In exercising our leadership, we will respect the values, judgment, and interests of our friends and partners. Still, we will be prepared to act apart when our interests and unique responsibilities require. When we disagree on particulars, we will explain forthrightly the grounds for our concerns and strive to forge viable alternatives. We will not allow such disagreements to obscure our determination to secure together, with our allies and our friends, our shared fundamental interests and values.

Ultimately, the foundation of American strength is at home. It is in the skills of our people, the dynamism of our economy, and the resilience of our institutions. A diverse, modern society has inherent, ambitious, entrepreneurial energy. Our strength comes from what we do with that energy. That is where our national security begins.

12

Joint Vision 2020

1. Introduction

The US military today is a force of superbly trained men and women who are ready to deliver victory for our Nation. In support of the objectives of our National Security Strategy, it is routinely employed to shape the international security environment and stands ready to respond across the full range of potential military operations. But the focus of this document is the third element of our strategic approach—the need to prepare now for an uncertain future.

Joint Vision 2020 builds upon and extends the conceptual template established by *Joint Vision 2010* to guide the continuing transformation of America's Armed Forces. The primary purpose of those forces has been and will be to fight and win the Nation's wars. The overall goal of the transformation described in this document is the creation of a force that is dominant across the full spectrum of military operations—persuasive in peace, decisive in war, preeminent in any form of conflict.

In 2020, the nation will face a wide range of interests, opportunities, and challenges and will require a military that can both win wars and contribute to peace. The global interests and responsibilities of the United States will endure, and there is no indication that threats to those interests and responsibilities, or to our allies, will disappear. The strategic concepts of decisive force, power projection, overseas presence, and strategic agility will continue to govern our efforts to fulfill those responsibilities and meet the challenges of the future. This document describes the operational concepts necessary to do so.

If our Armed Forces are to be faster, more lethal, and more precise in 2020 than they are today, we must continue to invest in and develop new military capabilities. This vision describes the ongoing transformation to those new capabilities. As first explained in *JV 2010,* and dependent upon realizing the potential of the information revolution, today's capabilities for maneuver, strike, logistics, and protection will become dominant maneuver, precision engagement, focused logistics, and full dimensional protection.

The joint force, because of its flexibility and responsiveness, will remain the key to operational success in the future. The integration of core competencies provided by the individual Services is essential to the joint team, and the employment of the capabilities of the Total Force (active, reserve, guard, and civilian members) increases the options for the commander and complicates the choices of our opponents. To build the most effective force for 2020, we must be fully joint: intellectually, operationally, organizationally, doctrinally, and technically.

This vision is centered on the joint force in 2020. The date defines a general analytical focus rather than serving as a definitive estimate or deadline. The document does not describe counters to specific threats, nor does it enumerate weapon, communication, or other systems we will develop or purchase. Rather, its purpose is to describe in broad terms the human talent—the professional, well-trained, and ready force—and operational capabilities that will be required for the joint force to succeed across the full range of military operations and accomplish its mission in 2020 and beyond. In describing those capabilities, the vision provides a vector for the wide-ranging program of exercises and experimentation being conducted by the Services and combatant commands and the continuing evolution of the joint force. Based on the joint vision implementation program, many capabilities will be operational well before 2020, while others will continue to be explored and developed through exercises and experimentation.

The overarching focus of this vision is full spectrum dominance—achieved through the interdependent application of dominant maneuver, precision engagement, focused logistics, and full dimensional protection. Attaining that goal requires the steady infusion of new technology and modernization and replacement of equipment. However, material superiority alone is not sufficient. Of greater importance is the development of doctrine, organizations, training and education, leaders, and people that effectively take advantage of the technology.

The evolution of these elements over the next two decades will be strongly influenced by two factors. First, the continued development and proliferation of information technologies will substantially change the conduct of military operations. These changes in the information environment make information superiority a key enabler of the transformation of the operational capabilities of the joint force and the evolution of joint command and control. Second, the US Armed Forces will continue to rely on a capacity for intellectual and technical innovation. The pace of technological change, especially as it fuels changes in the strategic environment, will place a premium on our ability to foster innovation in our people and organizations across the entire range of joint operations. The overall vision of the capabilities we will require in 2020, as introduced above, rests on our assessment of the strategic context in which our forces will operate.

2. Strategic Context

Three aspects of the world of 2020 have significant implications for the US Armed Forces. First, the United States will continue to have global interests and be engaged with a variety of regional actors. Transportation, communications, and information technology will continue to evolve and foster expanded economic ties and awareness of international events. Our security and economic interests, as well as our political values, will provide the impetus for engagement with international partners. The joint force of 2020 must be prepared to "win" across the full range of military operations in any part of the world, to operate with multinational forces, and to coordinate military operations, as necessary, with government agencies and international organizations.

Second, potential adversaries will have access to the global commercial industrial base and much of the same technology as the US military. We will not necessarily sustain a wide technological advantage over our adversaries in all areas. Increased availability of commercial satellites, digital communications, and the public internet all give adversaries new capabilities at a relatively low cost. We should not expect opponents in 2020 to fight with strictly "industrial age" tools. Our advantage must, therefore, come from leaders, people, doctrine, organizations, and training that enable us to take advantage of technology to achieve superior warfighting effectiveness.

Third, we should expect potential adversaries to adapt as our capabilities evolve. We have superior conventional warfighting capabilities and effective nuclear deterrence today, but this favorable military balance is not static. In the face of such strong capabilities, the appeal of asymmetric approaches and the focus on the development of niche capabilities will increase. By developing and using approaches that avoid US strengths and exploit potential vulnerabilities using significantly different methods of operation, adversaries will attempt to create conditions that effectively delay, deter, or counter the application of US military capabilities.

The potential of such asymmetric approaches is perhaps the most serious danger the United States faces in the immediate future—and this danger includes long-range ballistic missiles and other direct threats to US citizens and territory. The asymmetric methods and objectives of an adversary are often far more important than the relative technological imbalance, and the psychological impact of an attack might far outweigh the actual physical damage inflicted. An adversary may pursue an asymmetric advantage on the tactical, operational, or strategic level by identifying key vulnerabilities and devising asymmetric concepts and capabilities to strike or exploit them. To complicate matters, our adversaries may pursue a combination of asymmetries, or the United States may face a number of adversaries who, in combination, create an asymmetric threat. These asymmetric threats are dynamic and subject to change, and the US Armed Forces must maintain the capabilities necessary to deter, defend against, and defeat any adversary who chooses such an approach. To meet the challenges of the strategic environment in 2020, the joint force must be able to achieve full spectrum dominance.

3. Full Spectrum Dominance

The ultimate goal of our military force is to accomplish the objectives directed by the National Command Authorities. For the joint force of the future, this goal will be achieved through full spectrum dominance—the ability of US forces, operating unilaterally or in combination with multinational and interagency partners, to defeat any adversary and control any situation across the full range of military operations.

The full range of operations includes maintaining a posture of strategic deterrence. It includes theater engagement and presence activities. It includes conflict involving employment of strategic forces and weapons of mass destruction, major theater wars, regional conflicts, and smaller-scale contingencies. It also includes those ambiguous situations residing between peace and war, such as peacekeeping and peace enforcement operations, as well as noncombat humanitarian relief operations and support to domestic authorities.

The label full spectrum dominance implies that US forces are able to conduct prompt, sustained, and synchronized operations with combinations of forces tailored to specific situations and with access to and freedom to operate in all domains—space, sea, land, air, and information. Additionally, given the global nature of our interests and obligations, the United States must maintain its overseas presence forces and the ability to rapidly project power worldwide in order to achieve full spectrum dominance.

Achieving full spectrum dominance means the joint force will fulfill its primary purpose—victory in war, as well as achieving success across the full range of operations, but it does not mean that we will win without cost or difficulty. Conflict results in casualties despite our best efforts to minimize them, and will continue to do so when the force has achieved full spectrum dominance. Additionally, friction is inherent in military operations. The joint force of 2020 will seek to create a "frictional imbalance" in its favor by using the capabilities envisioned in this document, but the fundamental sources of friction cannot be eliminated. We will win—but we should not expect war in the future to be either easy or bloodless.

Sources of Friction

- Effects of danger and exertion
- Existence of uncertainty and chance
- Unpredictable actions of other actors
- Frailties of machines and information
- Humans

The requirement for global operations, the ability to counter adversaries who possess weapons of mass destruction, and the need to shape ambiguous situations at the low end of the range of operations will present special challenges en route to achieving full spectrum dominance. Therefore, the process of creating the joint force of the future must be flexible—to react to changes in the strategic environment and the adaptations of potential enemies, to take advantage of new technologies, and to account for variations in the pace of change. The source of that flexibility is the synergy of the core competencies of the individual Services, integrated into the joint team. These challenges will require a Total Force composed of well-educated, motivated, and competent people who can adapt to the many demands of future joint missions. The transformation of the joint force to reach full spectrum dominance rests upon information superiority as a key enabler and our capacity for innovation.

INFORMATION SUPERIORITY

Information environment—the aggregate of individuals, organizations, and systems that collect, process, or disseminate information, including the information itself. (JP1-02)

Information superiority—the capability to collect, process, and disseminate an uninterrupted flow of information while exploiting or denying an adversary's ability to do the same. (JP1-02) Information superiority is achieved in a noncombat situation or one in which there are no clearly defined adversaries when friendly forces have the information necessary to achieve operational objectives.

Information, information processing, and communications networks are at the core of every military activity. Throughout history, military leaders have regarded information superiority as a key enabler of victory. However, the ongoing "information revolution" is creating not only a quantitative, but a qualitative change in the information environment that by 2020 will result in profound changes in the conduct of military operations. In fact, advances in information capabilities are proceeding so rapidly that there is a risk of outstripping our ability to capture ideas, formulate operational concepts, and develop the capacity to assess results. While the goal of achieving information superiority will not change, the nature, scope, and "rules" of the quest are changing radically.

The qualitative change in the information environment extends the conceptual underpinnings of information superiority beyond the mere accumulation of more, or even better, information. The word "superiority" implies a state or condition of imbalance in one's favor. Information superiority is transitory in nature and must be created and sustained by the joint force through the conduct of information operations. However, the creation of information superiority is not an end in itself.

Information superiority provides the joint force a competitive advantage only when it is effectively translated into superior knowledge and decisions.

The joint force must be able to take advantage of superior information converted to superior knowledge to achieve "decision superiority" —better decisions arrived at and implemented faster than an opponent can react, or in a noncombat situation, at a tempo that allows the force to shape the situation or react to changes and accomplish its mission. Decision superiority does not automatically result from information superiority. Organizational and doctrinal adaptation, relevant training and experience, and the proper command and control mechanisms and tools are equally necessary.

The evolution of information technology will increasingly permit us to integrate the traditional forms of information operations with sophisticated all-source intelligence, surveillance, and reconnaissance in a fully synchronized information campaign. The development of a concept labeled the global information grid will provide the network-centric environment required to achieve this goal. The grid will be the globally interconnected, end-to-end set of information capabilities, associated processes, and people to manage and provide information on demand to warfighters, policy makers, and support personnel. It will enhance combat power and contribute to the success of noncombat military operations as well. Realization of the full potential of these changes requires not only technological improvements, but the continued evolution of organizations and doctrine and the development of relevant training to sustain a comparative advantage in the information environment.

We must also remember that information superiority neither equates to perfect information, nor does it mean the elimination of the fog of war. Information systems, processes, and operations add their own sources of friction and fog to the operational environment. Information superiority is fundamental to the transformation of the operational capabilities of the joint force. The joint force of 2020 will use superior information and knowledge to achieve decision superiority, to support advanced command and control capabilities, and to reach the full potential of dominant maneuver, precision engagement, full dimensional protection, and focused logistics. The breadth and pace of this evolution demands flexibility and a readiness to innovate.

INNOVATION

Joint Vision 2010 identified technological innovation as a vital component of the transformation of the joint force. Throughout the industrial age, the United States has relied upon its capacity for technological innovation to succeed in military operations, and the need to do so will continue. It is important, however, to broaden our focus beyond technology and capture the importance of organizational and conceptual innovation as well.

Innovation, in its simplest form, is the combination of new "things" with new "ways" to carry out tasks. In reality, it may result from fielding completely new things, or the imaginative recombination of old things in new ways, or something in between. The ideas in *JV 2010* as carried forward in JV 2020 are, indeed, innovative and form a vision for integrating doctrine, tactics, training, supporting activities, and technology into new operational capabilities. The innovations that determine joint and Service capabilities will result from a general understanding of what future conflict and military operations will be like, and a view of what the combatant commands and Services must do in order to accomplish assigned missions.

An effective innovation process requires continuous learning—a means of interaction and exchange that evaluates goals, operational lessons, exercises, experiments, and simulations—and that must include feedback mechanisms. The Services and combatant commands must allow our highly trained and skilled professionals the opportunity to create new concepts and ideas that may lead to future breakthroughs. We must foster the innovations necessary to create the joint force of the future—not only with decisions regarding future versus present force structure and budgets, but also with a reasonable tolerance for errors and failures in the experimentation process. We must be concerned with efficient use of time and resources and create a process that gives us confidence that our results will produce battlefield success. However, an experimentation process with a low tolerance for error makes it unlikely that the force will identify and nurture the most relevant and productive aspects of new concepts, capabilities, and technology. All individuals and organizations charged with experimentation in support of the evolution of our combat forces must ensure that our natural concern for husbanding resources and ultimately delivering successful results does not prevent us from pursuing innovations with dramatic if uncertain potential.

There is, of course, a high degree of uncertainty inherent in the pursuit of innovation. The key to coping with that uncertainty is bold leadership supported by as much information as possible. Leaders must assess the efficacy of new ideas, the potential drawbacks to new concepts, the capabilities of potential adversaries, the costs versus benefits of new technologies, and the organizational implications of new capabilities. They must make these assessments in the context of an evolving analysis of the economic, political, and technological factors of the anticipated security environment. Each of these assessments will have uncertainty associated with them. But the best innovations have often come from people who made decisions and achieved success despite uncertainties and limited information.

By creating innovation, the combatant commands and Services also create their best opportunities for coping with the increasing pace of change in the overall environment in which they function. Although changing technology is a primary driver of environmental change, it is not the only one. The search for innovation must encompass the entire context of joint operations—which means the Armed Forces must explore changes in doctrine, organization, training, materiel, leadership and education, personnel, and facilities as well as technology. Ultimately, the goal is to develop reasonable approaches with enough flexibility to recover from errors and unforeseen circumstances.

4. Conduct of Joint Operations

The complexities of the future security environment demand that the United States be prepared to face a wide range of threats of varying levels of intensity. Success in countering these threats will require the skillful integration of the core competencies of the Services into a joint force tailored to the specific situation and objectives. Commanders must be afforded the opportunity to achieve the level of effectiveness and synergy necessary to conduct decisive operations across the entire range of military operations. When combat operations are required, they must have an overwhelming array of capabilities available to conduct offensive and defensive operations and against which an enemy must defend. Other complex contingencies such as humanitarian relief or peace operations will require a rapid, flexible response to achieve national objectives in the required timeframe. Some situations may require the capabilities of only one Service, but in most cases, a joint force comprised of both Active and Reserve Components will be employed.

The complexity of future operations also requires that, in addition to operating jointly, our forces have the capability to participate effectively as one element of a unified national effort. This integrated approach brings to bear all the tools of statecraft to achieve our national objectives unilaterally when necessary, while making optimum use of the skills and resources provided by multinational military forces, regional and international organizations, non-governmental organizations, and private voluntary organizations when possible. Participation by the joint force in operations supporting civil authorities will also likely increase in importance due to emerging threats to the US homeland such as terrorism and weapons of mass destruction.

PEOPLE

The core of the joint force of 2020 will continue to be an All Volunteer Force composed of individuals of exceptional dedication and ability. Their quality will matter as never before as our Service members confront a diversity of missions and technological demands that call for adaptability, innovation, precise judgment, forward thinking, and multicultural understanding. The Nation will continue to depend on talented individuals of outstanding character, committed to an ethic of selfless service.

Our people will require a multitude of skills. The Services will play a critical role in perfecting their individual specialties and the core competencies of each organization. In addition, every member of the Total Force must be prepared to apply that expertise to a wide range of missions as a member of the joint team. Our Service members must have the mental agility to transition from preparing for war to enforcing peace to actual combat, when necessary. The joint force commander is thereby provided a powerful, synergistic force capable of dominating across the entire range of operations.

The missions of 2020 will demand Service members who can create and then take advantage of intellectual and technological innovations. Individuals will be challenged by significant responsibilities at tactical levels in the organization and must be capable of making decisions with both operational and strategic implications. Our vision of full spectrum dominance and the transformation of operational capabilities has significant implications for the training and education of our people. The tactics of information operations, the coordination of interagency and multinational operations, as well as the complexity of the modern tools of war all require people who are both talented and trained to exacting standards. Rapid and dispersed operations will require men and women who are part of a cohesive team, yet are capable of operating independently to meet the commander's intent. The evolution of new functional areas, such as space operations and information operations, will require development of appropriate career progression and leadership opportunities for specialists in those fields. The accumulation of training and experience will create a force ready to deploy rapidly to any point on the globe and operate effectively.

The joint force of 2020 will face a number of challenges in recruiting and retaining the outstanding people needed to meet these requirements. First, expanding civilian education and employment opportunities will reduce the number of candidates available for military service. We will continue to focus on our members' standard of living and a competitive compensation strategy to ensure we attract the quality individuals we need. Second, the increasing percentage of members with dependents will require a commitment to family-oriented community support programs and as much stability as possible, as well as close monitoring of the impact of the operations tempo. Finally, our increased dependence on the Reserve Component will require us to address the concerns of our reserve members and their employers regarding the impact on civilian careers. The Department of Defense and Services must meet these challenges head-on.

Military operations will continue to demand extraordinary dedication and sacrifice under the most adverse conditions. Our Total Force, composed of professionals armed with courage, stamina, and intellect, will succeed despite the complexity and pace of future operations.

INTEROPERABILITY

Interoperability—the ability of systems, units, or forces to provide services to and accept services from other systems, units, or forces and to use the services so exchanged to enable them to operate effectively together. (JP1-02)

Interoperability is the foundation of effective joint, multinational, and interagency operations. The joint force has made significant progress toward achieving an optimum level of interoperability, but there must be a concerted effort toward continued improvement. Further improvements will include the refinement of joint doctrine as well as further development of common technologies and processes. Exercises, personnel exchanges, agreement on standardized operating procedures, individual training and education, and planning will further enhance and institutionalize these capabilities. Interoperability is a mandate for the joint force of 2020—especially in terms of communications, common logistics items, and information sharing. Information systems and equipment that enable a common relevant operational picture must work from shared networks that can be accessed by any appropriately cleared participant.

Although technical interoperability is essential, it is not sufficient to ensure effective operations. There must be a suitable focus on procedural and organizational elements, and decision makers at all levels must understand each other's capabilities and constraints. Training and education, experience and exercises, cooperative planning, and skilled liaison at all levels of the joint force will not only overcome the barriers of organizational culture and differing priorities, but will teach members of the joint team to appreciate the full range of Service capabilities available to them.

The future joint force will have the embedded technologies and adaptive organizational structures that will allow trained and experienced people to develop compatible processes and procedures, engage in collaborative planning, and adapt as necessary to specific crisis situations. These features are not only vital to the joint force, but to multinational and interagency operations as well.

Multinational Operations

Multinational Operations—a collective term to describe military actions conducted by forces of two or more nations usually undertaken within the structure of a coalition or alliance. (JP1-02)

Since our potential multinational partners will have varying levels of technology, a tailored approach to interoperability that accommodates a wide range of needs and capabilities is necessary. Our more technically advanced allies will have systems and equipment that are essentially compatible, enabling them to interface and share information in order to operate effectively with US forces at all levels. However, we must also be capable of operating with allies and coalition partners who may be technologically incompatible—especially at the tactical level. Additionally, many of our future partners will have significant specialized capabilities that may be integrated into a common operating scheme. At the same time, the existence of these relationships does not imply access to information without constraints. We and our multinational partners will continue to use suitable judgment regarding the protection of sensitive information and information sources.

In all cases, effective command and control is the primary means of successfully extending the joint vision to multinational operations. Technological developments that connect the information systems of partners will provide the links that lead to a common relevant operational picture and improve command and control. However, the sharing of information needed to maintain the tempo of integrated multinational operations also relies heavily on a shared understanding of operational procedures and compatible organizations. The commander must have the ability to evaluate information in its multinational context. That context can only be appreciated if sufficient regional expertise and liaison capability are available on the commander's staff. A deep understanding of the cultural, political, military, and economic characteristics of a region must be established and maintained. Developing this understanding is dependent upon shared training and education, especially with key partners, and may require organizational change as well. The overall effectiveness of multinational operations is, therefore, dependent on interoperability between organizations, processes, and technologies.

Interagency Operations

Interagency Coordination—within the context of Department of Defense involvement, the coordination that occurs between elements of the Department of Defense and engaged US Government agencies, nongovernmental organizations, private voluntary organizations, and regional and international organizations for the purpose of accomplishing an objective. (JP1-02)

The primary challenge of interagency operations is to achieve unity of effort despite the diverse cultures, competing interests, and differing priorities of the participating organizations, many of whom guard their relative independence, freedom of action, and impartiality. Additionally, these organizations may lack the structure and resources to support extensive liaison cells or integrative technology. In this environment and in the absence of formal command relationships, the future joint force must be proactive in improving communications, planning, interoperability, and liaison with potential interagency participants. These factors are important in all aspects of interagency operations, but particularly in the context of direct threats to citizens and facilities in the US homeland. Cohesive interagency action is vital to deterring, defending against, and responding to such attacks. The joint force must be prepared to support civilian authorities in a fully integrated effort to meet the needs of US citizens and accomplish the objectives specified by the National Command Authorities.

All organizations have unique information assets that can contribute to the common relevant operational picture and support unified action. They also have unique information requirements. Sharing information with appropriately cleared participants and integration of information from all sources are essential. Understanding each other's requirements and assets is also crucial. More importantly, through training with potential interagency partners, experienced liaisons must be developed to support long-term relationships, collaborative planning in advance of crises, and compatible processes and procedures. As with our multinational partners, interoperability in all areas of interaction is essential to effective interagency operations.

OPERATIONAL CONCEPTS

Dominant Maneuver

Dominant Maneuver is the ability of joint forces to gain positional advantage with decisive speed and overwhelming operational tempo in the achievement of assigned military tasks. Widely dispersed joint air, land, sea, amphibious, special operations and space forces, capable of scaling and massing force or forces and the effects of fires as required for either combat or noncombat operations, will secure advantage across the range of military operations through the application of information, deception, engagement, mobility and counter-mobility capabilities.

The joint force capable of dominant maneuver will possess unmatched speed and agility in positioning and repositioning tailored forces from widely dispersed locations to achieve operational objectives quickly and decisively. The employment of dominant maneuver may lead to achieving objectives directly, but can also facilitate employment of the other operational concepts. For example, dominant maneuver may be employed to dislodge enemy forces so they can be destroyed through precision engagement. At times, achieving positional advantage will be a function of operational maneuver over strategic distances. Overseas or US-based units will mass forces or effects directly to the operational theater. Information superiority will support the conduct of dominant maneuver by enabling adaptive and concurrent planning; coordination of widely dispersed units; gathering of timely feedback on the status, location, and activities of subordinate units; and anticipation of the course of events leading to mission accomplishment. The joint force will also be capable of planning and conducting dominant maneuver in cooperation with interagency and multinational partners with varying levels of commitment and capability.

The capability to rapidly mass force or forces and the effects of dispersed forces allows the joint force commander to establish control of the battlespace at the proper time and place. In a conflict, this ability to attain positional advantage allows the commander to employ decisive combat power that will compel an adversary to react from a position of disadvantage, or quit. In other situations, it allows the force to occupy key positions to shape the course of events and minimize hostilities or react decisively if hostilities erupt. And in peacetime, it constitutes a credible capability that influences potential adversaries while reassuring friends and allies.

Beyond the actual physical presence of the force, dominant maneuver creates an impact in the minds of opponents and others in the operational area. That impact is a tool available to the joint force commander across the full range of military operations. In a conflict, for example, the presence or anticipated presence of a decisive force might well cause an enemy to surrender after minimal resistance. During a peacekeeping mission, it may provide motivation for good-faith negotiations or prevent the instigation of civil disturbances. In order to achieve such an impact, the commander will use information operations as a force multiplier by making the available combat power apparent without the need to physically move elements of the force. The joint force commander will be able to take advantage of the potential and actual effects of dominant maneuver to gain the greatest benefit.

Precision Engagement

Precision Engagement is the ability of joint forces to locate, surveil, discern, and track objectives or targets; select, organize, and use the correct systems; generate desired effects; assess results; and reengage with decisive speed and overwhelming operational tempo as required, throughout the full range of military operations.

Simply put, precision engagement is effects-based engagement that is relevant to all types of operations. Its success depends on in-depth analysis to identify and locate critical nodes and targets. The pivotal characteristic of precision engagement is the linking of sensors, delivery systems, and effects. In the joint force of the future, this linkage will take place across Services and will incorporate the applicable capabilities of multinational and interagency partners when appropriate. The resulting system of systems will provide the commander the broadest possible range of capabilities in responding to any situation, including both kinetic and nonkinetic weapons capable of creating the desired lethal or nonlethal effects.

The concept of precision engagement extends beyond precisely striking a target with explosive ordnance. Information superiority will enhance the capability of the joint force commander to understand the situation, determine the effects desired, select a course of action and the forces to execute it, accurately assess the effects of that action, and reengage as necessary while minimizing collateral damage. During conflict, the commander will use precision engagement to obtain lethal and nonlethal effects in support of the objectives of the campaign. This action could include destroying a target using conventional forces, inserting a special operations team, or even the execution of a comprehensive psychological operations mission. In other cases, precision engagement may be used to facilitate dominant maneuver and decisive close combat. The commander may also employ nonkinetic weapons, particularly in the arena of information operations where the targets might be key enemy leaders or troop formations, or the opinion of an adversary population.

In noncombat situations, precision engagement activities will, naturally, focus on nonlethal actions. These actions will be capable of defusing volatile situations, overcoming misinformation campaigns, or directing a flow of refugees to relief stations, for example. Regardless of its application in combat or noncombat operations, the capability to engage precisely allows the commander to shape the situation or battle space in order to achieve the desired effects while minimizing risk to friendly forces and contributing to the most effective use of resources.

Focused Logistics

Focused Logistics is the ability to provide the joint force the right personnel, equipment, and supplies in the right place, at the right time, and in the right quantity, across the full range of military operations. This will be made possible through a real-time, web-based information system providing total asset visibility as part of a common relevant operational picture, effectively linking the operator and logistician across Services and support agencies. Through transformational innovations to organizations and processes, focused logistics will provide the joint warfighter with support for all functions.

Focused logistics will provide military capability by ensuring delivery of the right equipment, supplies, and personnel in the right quantities, to the right place, at the right time to support operational objectives. It will result from revolutionary improvements in information systems, innovation in organizational structures, reengineered processes, and advances in transportation technologies. This transformation has already begun with changes scheduled for the near term (see events highlighted in box at right) facilitating the ultimate realization of the full potential of focused logistics.

Focused Logistics Transformation Path

FY 01, implement systems to assess customer confidence from end to end of the logistics chain using customer wait time metric.

FY 02, implement time definite delivery capabilities using a simplified priority system driven by the customer's required delivery date.

FY 04, implement fixed and deployable automated identification technologies and information systems that provide accurate, actionable total asset visibility.

FY 04 for early deploying forces and **FY 06** for the remaining forces, implement a web-based, shared data environment to ensure the joint warfighters' ability to make timely and confident logistics decisions.

Focused logistics will effectively link all logistics functions and units through advanced information systems that integrate real-time total asset visibility with a common relevant operational picture. These systems will incorporate enhanced decision support tools that will improve analysis, planning, and anticipation of warfighter requirements. They will also provide a more seamless connection to the commercial sector to take advantage of applicable advanced business practices and commercial economies. Combining these capabilities with innovative organizational structures and processes will result in dramatically improved end-to-end management of the entire logistics system and provide precise real-time control of the logistics pipeline to support the joint force commander's priorities. The increased speed, capacity, and efficiency of advanced transportation systems will further improve deployment, distribution, and sustainment. Mutual support relationships and collaborative planning will enable optimum cooperation with multinational and interagency partners.

The result for the joint force of the future will be an improved link between operations and logistics resulting in precise time-definite delivery of assets to the warfighter. This substantially improved operational effectiveness and efficiency, combined with increasing warfighter confidence in these new capabilities, will concurrently reduce sustainment requirements and the vulnerability of logistics lines of communication, while appropriately sizing and potentially reducing the logistics footprint. The capability for focused logistics will effectively support the joint force in combat and provide the primary operational element in the delivery of humanitarian or disaster relief, or other activities across the range of military operations.

Full Dimensional Protection

Full Dimensional Protection is the ability of the joint force to protect its personnel and other assets required to decisively execute assigned tasks. Full dimensional protection is achieved through the tailored selection and application of multilayered active and passive measures, within the domains of air, land, sea, space, and information across the range of military operations with an acceptable level of risk.

Our military forces must be capable of conducting decisive operations despite our adversaries' use of a wide range of weapons (including weapons of mass destruction), the conduct of information operations or terrorist attacks, or the presence of asymmetric threats during any phase of these operations. Our people and the other military and nonmilitary assets needed for the successful conduct of operations must be protected wherever they are located—from deployment, to theater combat, to redeployment. Full dimensional protection exists when the joint force can decisively achieve its mission with an acceptable degree of risk in both the physical and information domains.

The capability for full dimensional protection incorporates a complete array of both combat and noncombat actions in offensive and defensive operations, enabled by information superiority. It will be based upon active and passive defensive measures, including theater missile defenses and possibly limited missile defense of the United States; offensive countermeasures; security procedures; antiterrorism measures; enhanced intelligence collection and assessments; emergency preparedness; heightened security awareness; and proactive engagement strategies. Additionally, it will extend beyond the immediate theater of operations to protect our reach-back, logistics, and key capabilities in other locations. There is a critical need for protection of the information content and systems vital for operational success, including increased vigilance in counterintelligence and information security. The joint force of 2020 will integrate protective capabilities from multinational and interagency partners when available and will respond to their requirements when possible. Commanders will thoroughly assess and manage risk as they apply protective measures to specific operations, ensuring an appropriate level of safety, compatible with other mission objectives, is provided for all assets.

The joint force commander will thereby be provided an integrated architecture for protection, which will effectively manage risk to the joint force and other assets, and leverage the contributions of all echelons of our forces and those of our multinational and interagency partners. The result will be improved freedom of action for friendly forces and better protection at all echelons.

INFORMATION OPERATIONS

Information operations - those actions taken to affect an adversary's information and information systems while defending one's own information and information systems. (JP1-02) Information operations also include actions taken in a noncombat or ambiguous situation to protect one's own information and information systems as well as those taken to influence target information and information systems.

The Variables of Information Operations

- Multidimensional definition and meaning of "information"—target, weapon, resource, or domain of operations
- Level of action and desired effect—tactical, operational, strategic, or combination
- Objective of operations—providing information, perception management, battlefield dominance, command and control warfare, systemic disruption, or systemic destruction
- Nature of situation—peace, crisis, or conflict

Information operations are essential to achieving full spectrum dominance. The joint force must be capable of conducting information operations, the purpose of which is to facilitate and protect US decision-making processes, and in a conflict, degrade those of an adversary. While activities and capabilities employed to conduct information operations are traditional functions of military forces, the pace of change in the information environment dictates that we expand this view and explore broader information operations strategies and concepts. We must recognize that "nontraditional" adversaries who engage in "nontraditional" conflict are of particular importance in the information domain. The United States itself and US forces around the world are subject to information attacks on a continuous basis regardless of the level and degree of engagement in other domains of operation. The perpetrators of such attacks are not limited to the traditional concept of a uniformed military adversary. Additionally, the actions associated with information operations are wide-ranging—from physical destruction to psychological operations to computer network defense. The task of integrating information operations with other joint force operations is complicated by the need to understand the many variables involved.

Our understanding of the interrelationships of these variables and their impact on military operations will determine the nature of information operations in 2020. The joint force commander will conduct information operations whether facing an adversary during a conflict or engaged in humanitarian relief operations. Such operations will be synchronized with those of multinational and interagency partners as the situation dictates. New offensive capabilities such as computer network attack techniques are evolving. Activities such as information assurance, computer network defense, and counterdeception will defend decision-making processes by neutralizing an adversaries' perception management and intelligence collection efforts, as well as direct attacks on our information systems. Because the ultimate target of information operations is the human decision maker, the joint force commander will have difficulty accurately assessing the effects of those operations. This problem of "battle damage assessment" for information operations is difficult and must be explored through exercises and rigorous experimentation.

The continuing evolution of information operations and the global information environment holds two significant implications. First, operations within the information domain will become as important as those conducted in the domains of sea, land, air, and space. Such operations will be inextricably linked to focused logistics, full dimensional protection, precision engagement, and dominant maneuver, as well as joint command and control. At the same time, information operations may evolve into a separate mission area requiring the Services to maintain appropriately designed organizations and trained specialists. Improvements in doctrine, organization, and technology may lead to decisive outcomes resulting primarily from information operations. As information operations continue to evolve, they, like other military operations, will be conducted consistent with the norms of our society, our alliances with other democratic states, and full respect for the laws of armed conflict. Second, there is significant potential for asymmetric engagements in the information domain. The United States has enjoyed a distinct technological advantage in the information environment and will likely continue to do so. However, as potential adversaries reap the benefits of the information revolution, the comparative advantage for the US and its partners will become more difficult to maintain. Additionally, our ever-increasing dependence on information processes, systems, and technologies adds potential vulnerabilities that must be defended.

JOINT COMMAND AND CONTROL

Command and control—the exercise of authority and direction by a properly designated commander over assigned and attached forces in the accomplishment of the mission. Command and control functions are performed through an arrangement of personnel, equipment, communications, facilities, and procedures employed by a commander in planning, directing, coordinating, and controlling forces and operations in the accomplishment of the mission. (JP1-02)

Command and control is the exercise of authority and direction over the joint force. It is necessary for the integration of the Services' core competencies into effective joint operations. The increasing importance of multinational and interagency aspects of the operations adds complexity and heightens the challenge of doing so. Command and control includes planning, directing, coordinating, and controlling forces and operations, and is focused on the effective execution of the operational plan; but the central function is decision making.

Command and control is most effective when decision superiority exists. Decision superiority results from superior information filtered through the commander's experience, knowledge, training, and judgment; the expertise of supporting staffs and other organizations; and the efficiency of associated processes. While changes in the information environment have led some to focus solely on the contribution of information superiority to command and control, it is equally necessary to understand the complete realm of command and control decision making, the nature of organizational collaboration, and especially, the "human in the loop."

In the joint force of the future, command and control will remain the primary integrating and coordinating function for operational capabilities and Service components. As the nature of military operations evolves, there is a need to evaluate continually the nature of command and control organizations, mechanisms, systems, and tools. There are two major issues to address in this evaluation—command structures and processes, and the information systems and technologies that are best suited to support them. Encompassed within these two issues, examination of the following related ideas and desired capabilities will serve as a catalyst for changes in doctrine, organization, and training.

- Commanders will need a broad understanding of new operational capabilities and new (often highly automated) supporting tools in order to be capable of flexible, adaptive coordination and direction of both forces and sensors.
- The staffs that support commanders must be organized and trained to take advantage of new capabilities. Commanders and staffs must also be capable of command and control in the face of technology failure.
- Commanders will be able to formulate and disseminate intent based upon up-to-date knowledge of the situation existing in the battlespace.
- Joint force headquarters will be dispersed and survivable and capable of coordinating dispersed units and operations. Subordinate headquarters will be small, agile, mobile, dispersed, and networked.
- Faster operations tempos, increased choices among weapons and effects, and greater weapons ranges will require continuous, simultaneous planning and execution at all levels.
- Expanding roles for multinational and interagency partners will require collaborative planning capabilities, technological compatibility/interoperability, and mechanisms for efficient information sharing.

Finally, as these and other changes take place over time, we must carefully examine three aspects of the human element of command and control. First, leaders of the joint force must analyze and understand the meaning of unit cohesion in the context of the small, widely dispersed units that are now envisioned. Second, decision makers at all levels must understand the implications of new technologies that operate continuously in all conditions when human beings are incapable of the same endurance. Third, as new information technologies, systems, and procedures make the same detailed information available at all levels of the chain of command, leaders must understand the implications for decision-making processes, the training of decision makers at all levels, and organizational patterns and procedures.

The potential for overcentralization of control and the capacity for relatively junior leaders to make decisions with strategic impact are of particular importance.

It has often been said that command is an art and control is a science—a basic truth that will remain true. Our thinking about command and control must be conceptually based, rather than focused on technology or materiel. Joint command and control is a nexus—a point of connection. It serves as a focal point for humans and technology, our evolving operational capabilities, and the capabilities of the Services. The development of effective joint command and control for the future requires rigorous and wide-ranging experimentation, focused especially on organizational innovation and doctrinal change.

5. Implementation

From Vision to Experimentation

- Joint Vision 2010 (1996)
- Concept for Future Joint Operations (1996–7)
- 21st Century Challenges and Desired Operational Capabilities (1997)
- Joint Warfighting Experimentation Program established, USACOM (JFCOM) as Executive Agent (1998)
- Joint Vision Implementation Master Plan (1998)
- CJCSI 3170, Requirements Generation System (1999)
- JFCOM Joint Experimentation Campaign Plans (1999 and 2000)

Joint Vision 2010, has had a profound impact on the development of US military capabilities. By describing those capabilities necessary to achieve success in 2010, we set in motion three important efforts. First, *JV 2010* established a common framework and language for the Services to develop and explain their unique contributions to the joint force. Second, we created a process for the conduct of joint experimentation and training to test ideas against practice. Finally, we began a process to manage the transformation of doctrine, organization, training, materiel, leadership and education, personnel, and facilities necessary to make the vision a reality. *Joint Vision 2020* builds on this foundation of success and will sustain the momentum of these processes.

The foundation of jointness is the strength of individual Service competencies pulled together. Our objective in implementing the joint vision is the optimal integration of all joint forces and effects. To achieve that goal, the interdependence of the Services requires mutual trust and reliance among all warfighters and a significantly improved level of interoperability—especially in the areas of command and control and sustainment. This interdependence will ultimately result in a whole greater than the sum of its parts, and will contribute to achieving full spectrum dominance through all forces acting in concert. The synergy gained through the interdependence of the Services makes clear that jointness is more than interoperability. The joint force requires capabilities that are beyond the simple combination of Service capabilities, and joint experimentation is the process by which those capabilities will be achieved.

To ensure unity of effort and continuity for joint concept development and experimentation, the Secretary of Defense designated the Commander in Chief, Joint Forces Command as the Executive Agent for experimentation design, preparation, execution, and assessment. Annual campaign plans provide focus to this effort and continuity in experimentation. The results of this iterative experimentation cycle are forwarded as comprehensive recommendations for changes in doctrine, organization, training, materiel, leadership and education, personnel, and facilities and lead to the co-evolution of all those elements. The experimentation and implementation process supporting the transformation of the joint force is also dependent upon Service and combatant command exercises and experimentation activities. The Service and combatant command visions support the joint vision by providing guidance for these individual efforts that are congruent with the Chairman's vision. Thus, in their own experimentation venues, the Services may develop recommendations with joint implications and will forward them to the appropriate joint experimentation activity.

To effect transforming and enduring changes to our joint military capabilities, the experimentation and implementation process must include construction of a wide range of scenarios and imaginative conflict simulations to explore the shape of future operations. Such intensive exploration of alternative concepts of operations can help the US military choose innovations that take the greatest advantage of combinations of new ideas and new technologies. The rapid pace of such changes will then drive further development of the experimentation and implementation process to field improved capabilities for the joint force.

The linchpin of progress from vision to experimentation to reality is joint training and education—because they are the keys to intellectual change. Without intellectual change, there is no real change in doctrine, organizations, or leaders. Thus, the implementation process is dependent upon incorporating concepts validated by experimentation into joint professional military education programs and joint exercises. In this way, individual Service members and units become a joint team capable of success across the full range of military operations.

6. Conclusion

This vision is firmly grounded in the view that the US military must be a joint force capable of full spectrum dominance. Its basis is four-fold: the global interests of the United States and the continuing existence of a wide range of potential threats to those interests; the centrality of information technology to the evolution of not only our own military, but also the capabilities of other actors around the globe; the premium a continuing broad range of military operations will place on the successful integration of multinational and interagency partners and the interoperability of processes, organizations, and systems; and our reliance on the joint force as the foundation of future US military operations.

Joint Vision 2020 builds on the foundation and maintains the momentum established with *Joint Vision 2010*. It confirms the direction of the ongoing transformation of operational capabilities, and emphasizes the importance of further experimentation, exercises, analysis, and conceptual thought, especially in the arenas of information operations, joint command and control, and multinational and interagency operations.

This vision recognizes the importance of technology and technical innovation to the US military and its operations. At the same time, it emphasizes that technological innovation must be accompanied by intellectual innovation leading to changes in organization and doctrine. Only then can we reach the full potential of the joint force—decisive capabilities across the full range of military operations. Such a vision depends upon the skill, experience, and training of the people comprising the Total Force and their leaders. The major innovations necessary to operate in the environment depicted herein can only be achieved through the recruitment, development, and retention of men and women with the courage, determination, and strength to ensure we are persuasive in peace, decisive in war, and preeminent in any form of conflict.

13

Sea Power 21 Series: Projecting Decisive Joint Capabilities

By Admiral Vern Clark, U.S. Navy
Proceedings, October 2002

Sea-based operations use revolutionary information superiority and dispersed, networked force capabilities to deliver unprecedented offensive power, defensive assurance, and operational independence to Joint Force Commanders.

Our Vision

The 21st century sets the stage for tremendous increases in naval precision, reach, and connectivity, ushering in a new era of joint operational effectiveness. Innovative concepts and technologies will integrate sea, land, air, space, and cyberspace to a greater extent than ever before. In this unified battlespace, the sea will provide a vast maneuver area from which to project direct and decisive power around the globe.

Future naval operations will use revolutionary information superiority and dispersed, networked force capabilities to deliver unprecedented offensive power, defensive assurance, and operational independence to

Joint Force Commanders. Our Navy and its partners will dominate the continuum of warfare from the maritime domain—deterring forward in peacetime, responding to crises, and fighting and winning wars.

By doing so, we will continue the evolution of U.S. naval power from the blue-water, war-at-sea focus of the "Maritime Strategy" (1986), through the littoral emphasis of ". . . From the Sea" (1992) and "Forward . . . from the Sea" (1994), to a broadened strategy in which naval forces are fully integrated into global joint operations against regional and transnational dangers.

To realize the opportunities and navigate the challenges ahead, we must have a clear vision of how our Navy will organize, integrate, and transform. "Sea Power 21" is that vision. It will align our efforts, accelerate our progress, and realize the potential of our people. "Sea Power 21" will guide our Navy as we defend our nation and defeat our enemies in the uncertain century before us.

- Sea Strike—Projecting Precise and Persistent Offensive Power
- Sea Shield—Projecting Global Defensive Assurance
- Sea Basing—Projecting Joint Operational Independence

Transformation for a Violent Era

The events of 11 September 2001 tragically illustrated that the promise of peace and security in the new century is fraught with profound dangers: nations poised for conflict in key regions, widely dispersed and well-funded terrorist and criminal organizations, and failed states that deliver only despair to their people.

These dangers will produce frequent crises, often with little warning of timing, size, location, or intensity. Associated threats will be varied and deadly, including weapons of mass destruction, conventional warfare, and widespread terrorism. Future enemies will attempt to deny us access to critical areas of the world, threaten vital friends and interests overseas, and even try to conduct further attacks against the American homeland. These threats will pose increasingly complex challenges to national security and future warfighting.

Previous strategies addressed regional challenges. Today, we must think more broadly. Enhancing security in this dynamic environment requires us to expand our strategic focus to include both evolving regional challenges and transnational threats. This combination of traditional and emerging dangers means increased risk to our nation. To counter that risk, our Navy must expand its striking power, achieve information dominance, and develop transformational ways of fulfilling our enduring missions of sea control, power projection, strategic deterrence, strategic sealift, and forward presence.

Three fundamental concepts lie at the heart of the Navy's continued operational effectiveness: Sea Strike, Sea Shield, and Sea Basing. Sea Strike is the ability to project precise and persistent offensive power from the sea; Sea Shield extends defensive assurance throughout the world; and Sea Basing enhances operational independence and support for the joint force. These concepts build upon the solid foundation of the Navy-Marine Corps team, leverage U.S. asymmetric advantages, and strengthen joint combat effectiveness.

We often cite asymmetric challenges when referring to enemy threats, virtually assuming such advantages belong only to our adversaries. "Sea Power 21" is built on a foundation of American asymmetric strengths that are powerful and uniquely ours. Among others, these include the expanding power of computing, systems integration, a thriving industrial base, and the extraordinary capabilities of our people, whose innovative nature and desire to excel give us our greatest competitive advantage.

Sea Strike, Sea Shield, and Sea Basing will be enabled by ForceNet, an overarching effort to integrate warriors, sensors, networks, command and control, platforms, and weapons into a fully netted, combat force. We have been talking about network-centric warfare for a decade, and ForceNet will be the Navy's plan to make it an operational reality. Supported by ForceNet, Sea Strike, Sea Shield, and Sea Basing capabilities will be deployed by way of a Global Concept of Operations that widely distributes the firepower of the fleet, strengthens deterrence, improves crisis response, and positions us to win decisively in war.

Sea Strike: Projecting Precise and Persistent Offensive Power

Projecting decisive combat power has been critical to every commander who ever went into battle, and this will remain true in decades ahead. Sea Strike operations are how the 21st-century Navy will exert direct, decisive, and sustained influence in joint campaigns. They will involve the dynamic application of persistent intelligence, surveillance, and reconnaissance; time-sensitive strike; ship-to-objective maneuver; information operations; and covert strike to deliver devastating power and accuracy in future campaigns.

Information gathering and management are at the heart of this revolution in striking power. Networked, long-dwell naval sensors will be integrated with national and joint systems to penetrate all types of cover and weather, assembling vast amounts of information. Data provided by Navy assets—manned and unmanned—will be vital to establishing a comprehensive understanding of enemy military, economic, and political vulnerabilities. Rapid planning processes will then use this knowledge to tailor joint strike packages that deliver calibrated effects at precise times and places.

Sea Strike Impact

- Amplified, effects-based striking power
- Increased precision attack and information operations
- Enhanced warfighting contribution of Marines and Special Forces
- "24 / 7" offensive operations
- Seamless integration with joint strike packages

Sea Strike Capabilities

- Persistent intelligence, surveillance, and reconnaissance
- Time-sensitive strike
- Electronic warfare/information operations
- Ship-to-objective maneuver
- Covert strike

Future Sea Strike Technologies

- Autonomous, organic, long-dwell sensors
- Integrated national, theater, and force sensors
- Knowledge-enhancement systems
- Unmanned combat vehicles
- Hypersonic missiles
- Electro-magnetic rail guns
- Hyper-spectral imaging

Sea Strike: Action Steps

- Accelerate information dominance via ForceNet
- Develop, acquire, and integrate systems to increase combat reach, stealth, and lethality
- Distribute offensive striking capability throughout the entire force
- Deploy sea-based, long-dwell, manned and unmanned sensors

- Develop information operations as a major warfare area
- Synergize with Marine Corps transformation efforts
- Partner with the other services to accelerate Navy transformation

Knowledge dominance provided by persistent intelligence, surveillance, and reconnaissance will be converted into action by a full array of Sea Strike options—next-generation missiles capable of in-flight targeting, aircraft with stand-off precision weapons, extended-range naval gunfire, information operations, stealthy submarines, unmanned combat vehicles, and Marines and SEALs on the ground. Sovereign naval forces will exploit their strategic flexibility, operational independence, and speed of command to conduct sustained operations 24 hours per day, 7 days per week, 365 days per year.

Information superiority and flexible strike options will result in time-sensitive targeting with far greater speed and accuracy. Military operations will become more complicated as advanced intelligence, surveillance, and reconnaissance products proliferate. Expanded situational awareness will put massed forces at risk, for both friends and adversaries. This will compress timelines and prompt greater use of dispersed, low-visibility forces. Countering such forces will demand speed, agility, and streamlined information processing tied to precision attack. Sea Strike will meet that challenge.

The importance of information operations will grow in the years ahead as high-technology weapons and systems become more widely available. Information operations will mature into a major warfare area, to include electronic warfare, psychological operations, computer network attack, computer network defense, operations security, and military deception. Information operations will play a key role in controlling crisis escalation and preparing the battlefield for subsequent attack. This U.S. asymmetry will be a critical part of Sea Strike.

When we cannot achieve operational objectives from over the horizon, our Navy-Marine Corps team moves ashore. Using advanced vertical and horizontal envelopment techniques, fully netted ground forces will maneuver throughout the battlespace, employing speed and precision to generate combat power. Supported by sea bases, we will exploit superior situational awareness and coordinated fires to create shock, confusion, and chaos in enemy ranks. Information superiority and networking will act as force multipliers, allowing agile ground units to produce the warfighting impact traditionally provided by far heavier forces, bringing expeditionary warfare to a new level of lethality and combat effectiveness.

Sea Strike capabilities will provide Joint Force Commanders with a potent mix of weapons, ranging from long-range precision strike, to covert land—attack in anti-access environments, to the swift insertion of ground forces. Information superiority will empower us to dominate time-lines, foreclose adversary options, and deny enemy sanctuary. Sea Strike operations will be fully integrated into joint campaigns, adding the unique independence, responsiveness, and on-scene endurance of naval forces to joint strike efforts. Combined sea-based and land-based striking power will produce devastating effects against enemy strategic, operational, and tactical pressure points-resulting in rapid, decisive operations and the early termination of conflict.

Sea Shield: Projecting Global Defensive Assurance

Traditionally, naval defense has protected the unit, the fleet, and the sea lines of communication. Tomorrow's Navy will do much more. Sea Shield takes us beyond unit and task-force defense to provide the nation with sea-based theater and strategic defense.

Sea Shield will protect our national interests with layered global defensive power based on control of the seas, forward presence, and networked intelligence. It will use these strengths to enhance homeland defense, assure access to contested littorals, and project defensive power deep inland. As with Sea Strike, the foundation of these integrated operations will be information superiority, total force networking, and an agile and flexible sea-based force.

Homeland defense will be accomplished by a national effort that integrates forward-deployed naval forces with the other military services, civil authorities, and intelligence and law-enforcement agencies. Working with the newly established Northern Command, we will identify, track, and intercept dangers long before they threaten our homeland. These operations will extend the security of the United States far seaward, taking advantage of the time and space afforded by naval forces to shield our nation from impending threats.

Sea Shield Impact

- Projected defense for joint forces and allies ashore
- Sustained access for maritime trade, coalition building, and military operations

- Extended homeland defense via forward presence and networked intelligence
- Enhanced international stability, security, and engagement

Sea Shield Capabilities

- Homeland defense
- Sea/littoral superiority
- Theater air missile defense
- Force entry enabling

Future Sea Shield Technologies

- Interagency intelligence and communications reach-back systems
- Organic mine countermeasures
- Multi-sensor cargo inspection equipment
- Advanced hull forms and modular mission payloads
- Directed-energy weapons
- Autonomous unmanned vehicles
- Common undersea picture
- Single integrated air picture
- Distributed weapons coordination
- Theater missile defense

Sea Shield: Action Steps

- Expand combat reach
- Deploy theater missile defense as soon as possible
- Create common operational pictures for air, surface, and subsurface forces
- Accelerate the development of sea-based unmanned vehicles to operate in every environment
- Invest in self-defense capabilities to ensure sea superiority

Maritime patrol aircraft, ships, submarines, and unmanned vehicles will provide comprehensive situational awareness to cue intercepting units. When sent to investigate a suspicious vessel, boarding parties will use advanced equipment to detect the presence of contraband by visual, chemical, and radiological methods. Forward-deployed naval forces will also protect the homeland by engaging inbound ballistic missiles in the boost or mid-course phase, when they are most vulnerable to interception. In addition, our nuclear-armed Trident ballistic missile submarine force will remain on silent patrol around the world, providing the ultimate measure of strategic deterrence. These highly survivable submarines are uniquely powerful assets for deterring aggressors who would contemplate using weapons of mass destruction.

Achieving battle-space superiority in forward theaters is central to the Sea Shield concept, especially as enemy area-denial efforts become more capable. In times of rising tension, pre-positioned naval units will sustain access for friendly forces and maritime trade by employing evolving expeditionary sensor grids and advanced deployable systems to locate and track enemy threats. Speed will be an ally as linked sensors, high-speed platforms, and improved kill vehicles consolidate area control, including the location and neutralization of mines via state-of-the-art technology on dedicated mine warfare platforms and battle group combatants. Mission-reconfigurable Littoral Combat Ships, manned and unmanned aviation assets, and submarines with unmanned underwater vehicles will gain and maintain the operational advantage, while sea-based aircraft and missiles deliver air dominance. The result will be combat-ready forces that are prepared to "climb into the ring" to achieve and sustain access before and during crises.

Perhaps the most dramatic advancement promised by Sea Shield will be the ability of naval forces to project defensive power deep overland, assuring friends and allies while protecting joint forces ashore. A next-generation long-range surface-to-air Standard Missile, modernized E-2 Hawkeye radar, and Cooperative Engagement Capability will combine to extend sea-based cruise missile defense far inland. This will reinforce the impact of sea-based ballistic missile defense and greatly expand the coverage of naval area defense. These capabilities represent a broadened mission for our Navy that will lessen the defensive burden on land forces and increase sea-based influence over operations ashore.

The importance of Sea Shield to our nation has never been greater, as the proliferation of advanced weapons and asymmetric attack techniques places an increasing premium on the value of deterrence and battlespace dominance. Sea Shield capabilities, deployed forward, will help dissuade aggressors before the onset of conflict. In addition, Sea Shield will complement Sea Strike efforts by freeing aviation forces previously devoted to force defense, allowing them to concentrate on strike missions and generate far greater offensive firepower from the fleet. In sum, Sea Shield will enhance crisis control, protect allies and joint forces ashore, and set the stage for combat victory—providing a powerful new tool for joint combatant commanders in this dangerous age.

Sea Basing: Projecting Joint Operational Independence

Operational maneuver is now, and always has been, fundamental to military success. As we look to the future, the extended reach of networked weapons and sensors will tremendously increase the impact of naval forces in joint campaigns. We will do this by exploiting the largest maneuver area on the face of the earth: the sea.

Sea Basing serves as the foundation from which offensive and defensive fires are projected—making Sea Strike and Sea Shield realities. As enemy access to weapons of mass destruction grows, and the availability of overseas bases declines, it is compelling both militarily and politically to reduce the vulnerability of U.S. forces through expanded use of secure, mobile, networked sea bases. Sea Basing capabilities will include providing Joint Force Commanders with global command and control and extending integrated logistical support to other services. Afloat positioning of these capabilities strengthens force protection and frees airlift-sealift to support missions ashore.

Sea Basing Impact

- Pre-positioned warfighting capabilities for immediate employment
- Enhanced joint support from a fully netted, dispersed naval force
- Strengthened international coalition building
- Increased joint force security and operational agility
- Minimized operational reliance on shore infrastructure

Sea Basing Capabilities

- Enhanced afloat positioning of joint assets
- Offensive and defensive power projection
- Command and control
- Integrated joint logistics
- Accelerated deployment and employment timelines

Future Sea Basing Technologies

- Enhanced sea-based joint command and control
- Heavy equipment transfer capabilities

- Intra-theater high-speed sealift
- Improved vertical delivery methods
- Integrated joint logistics
- Rotational crewing infrastructure
- International data-sharing networks

Sea Basing: Action Steps

- Exploit the advantages of sea-based forces wherever possible
- Develop technologies to enhance on-station time and minimize maintenance requirements
- Experiment with innovative employment concepts and platforms
- Challenge every assumption that results in shore basing of Navy capabilities

Netted and dispersed sea bases will consist of numerous platforms, including nuclear-powered aircraft carriers, multi-mission destroyers, submarines with Special Forces, and maritime pre-positioned ships, providing greatly expanded power to joint operations. Sea-based platforms will also enhance coalition-building efforts, sharing their information and combat effectiveness with other nations in times of crisis.

Sea Basing accelerates expeditionary deployment and employment timelines by pre-positioning vital equipment and supplies in-theater, preparing the United States to take swift and decisive action during crises. We intend to develop these capabilities to the fullest extent. Strategic sealift will be central to this effort. It remains a primary mission of the U.S. Navy and will be critical during any large conflict fought ashore. Moreover, we will build pre-positioned ships with at-sea-accessible cargo, awaiting closure of troops by way of high-speed sealift and airlift. Joint operational flexibility will be greatly enhanced by employing pre-positioned shipping that does not have to enter port to offload.

Twenty-first-century operations will require greater efficiencies through the development of joint logistical support. This will include the provisioning of joint supplies and common ammunition, and the completion of critical repairs from afloat platforms. Providing these capabilities to on-scene commanders will significantly increase operational effectiveness and constitute a valuable addition to strategic basing support provided by friends and allies around the world.

Beyond its operational impact, the Sea Basing concept provides a valuable tool for prioritizing naval programs. Sea-based forces enjoy advantages of security, immediate employability, and operational independence. All naval programs should foster these attributes to the greatest extent feasible. This means transforming shore-based capabilities to sea-based systems whenever practical, and improving the reach, persistence, and sustainability of systems that are already afloat.

ForceNet: Enabling 21st Century Warfare

ForceNet is the "glue" that binds together Sea Strike, Sea Shield, and Sea Basing. It is the operational construct and architectural framework for naval warfare in the information age, integrating warriors, sensors, command and control, platforms, and weapons into a networked, distributed combat force.

ForceNet will provide the architecture to increase substantially combat capabilities through aligned and integrated systems, functions, and missions. It will transform situational awareness, accelerate speed of decision, and allow us to greatly distribute combat power. ForceNet will harness information for knowledge-based combat operations and increase force survivability. It will also provide real-time enhanced collaborative planning among joint and coalition partners.

ForceNet Impact

- Connected warriors, sensors, networks, command and control, platforms, and weapons
- Accelerated speed and accuracy of decision
- Integrated knowledge to dominate the battlespace

ForceNet Capabilities

- Expeditionary, multi-tiered, sensor and weapons grids
- Distributed, collaborative command and control
- Dynamic, multi-path and survivable networks
- Adaptive/automated decision aids
- Human-centric integration

Using a total system approach, ForceNet will shape the development of integrated capabilities. These include maritime information processing and command and control components that are fully interoperable with joint systems; intelligence, surveillance, and reconnaissance fusion capabilities to support rapid targeting and maneuver; open systems architecture for broad and affordable interoperability; and safeguards to ensure networks are reliable and survivable. ForceNet also emphasizes the human factor in the development of advanced technologies. This philosophy acknowledges that the warrior is a premier element of all operational systems.

Today, ForceNet is moving from concept to reality. Initial efforts will focus on integrating existing networks, sensors, and command and control systems. In the years ahead, it will enable the naval service to employ a fully netted force, engage with distributed combat power, and command with increased awareness and speed as an integral part of the joint team.

Global Concept of Operations

"Sea Power 21" will be implemented by a Global Concept of Operations that will provide our nation with widely dispersed combat power from platforms possessing unprecedented warfighting capabilities. The global environment and our defense strategy call for a military with the ability to respond swiftly to a broad range of scenarios and defend the vital interests of the United States. We must dissuade, deter, and defeat both regional adversaries and transnational threats.

The Global Concept of Operations will disperse combat striking power by creating additional independent operational groups capable of responding simultaneously around the world. This increase of combat power is possible because technological advancements are dramatically transforming the capability of our ships, submarines, and aircraft to act as power projection forces, netted together for expanded warfighting effect.

Impact of Global Concept of Operations

- Widely distributed, fully netted striking power to support joint operations
- Increased presence, enhanced flexibility, and improved responsiveness
- Task-organized to deter forward, respond to crises, and win decisively

The results will be profound. Naval capability packages will be readily assembled from forward-deployed forces. These forces will be tailored to meet the mission needs of the Joint Force Commander, complementing other available joint assets. They will be sized to the magnitude of the task at hand. As a result, our Navy will be able to respond simultaneously to a broad continuum of contingencies and conflict, anywhere around the world. The Global Concept of Operations will employ a flexible force structure that includes:

- Carrier Strike Groups that provide the full range of operational capabilities. Carrier Strike Groups will remain the core of our Navy's warfighting strength. No other force package will come close to matching their sustained power projection ability, extended situational awareness, and combat survivability.
- Expeditionary Strike Groups consisting of amphibious ready groups augmented with strike-capable surface warships and submarines. These groups will prosecute Sea Strike missions in lesser-threat environments. As our operational concepts evolve, and new systems like Joint Strike Fighter deliver to the fleet, it will be advantageous to maximize this increased aviation capability. New platforms being developed for Expeditionary Strike Groups should be designed to realize this warfighting potential.
- Missile-defense Surface Action Groups will increase international stability by providing security to allies and joint forces ashore.
- Specially modified Trident submarines will provide covert striking power from cruise missiles and the insertion of Special Operations Forces.
- A modern, enhanced-capability Combat Logistics Force will sustain the widely dispersed fleet.

The Global Concept of Operations requires a fleet of approximately 375 ships that will increase our striking power from today's 12 carrier battle groups, to 12 Carrier Strike Groups, 12 Expeditionary Strike Groups, and multiple missile-defense Surface Action Groups and guided-missile submarines. These groups will operate independently around the world to counter transnational threats and they will join together to form Expeditionary Strike Forces—the "gold standard" of naval power—when engaged in regional conflict.

This dispersed, netted, and operationally agile fleet, as part of the joint force, will deliver the combat power needed to sustain homeland defense, provide forward deterrence in four theaters, swiftly defeat two aggressors at the same time, and deliver decisive victory in one of those conflicts. Employment of sovereign sea-based forces projecting offensive and defensive power across a unified battlespace will be central to every war plan. Equally important, this 21st-century fleet will be positioned to immediately counter unexpected threats arising from any corner of the world.

The Global Concept of Operations will increase striking power, enhance flexibility, and improve responsiveness. It will fulfill our broadened strategy by sustaining the on-scene capabilities needed to fight and win.

Achieving Our Vision

We are developing Sea Strike, Sea Shield, and Sea Basing through a supporting triad of organizational processes: Sea Trial, Sea Warrior, and Sea Enterprise—initiatives that will align and accelerate the development of enhanced warfighting capabilities for the fleet.

Sea Trial: The Process of Innovation

Our enemies are dedicated to finding new and effective methods of attacking us. They will not stand still. To outpace our adversaries, we must implement a continual process of rapid concept and technology development that will deliver enhanced capabilities to our Sailors as swiftly as possible.

The Navy starts with the fleet, and Sea Trial will be fleet-led. The Commander, U.S. Fleet Forces Command, will serve as Executive Agent for Sea Trial, with Second and Third Fleet commanders sponsoring the development of Sea Strike, Sea Shield, and Sea Basing capabilities. These commanders will reach throughout the military and beyond to coordinate concept and technology development in support of future warfighting effectiveness. The Systems Commands and Program Executive Offices will be integral partners in this effort, bringing concepts to reality through technology innovation and the application of sound business principles.

The Navy Warfare Development Command, reporting directly to the Commander, U.S. Fleet Forces Command, will coordinate Sea Trial. Working closely with the fleets, technology development centers, and academic resources, the Navy Warfare Development Command will integrate wargaming, experimentation, and exercises to speed development of new concepts and technologies. They will do this by identifying candidates with the greatest potential to provide dramatic increases in warfighting capability. Embracing spiral development, these technologies and concepts will then be matured through targeted investment and guided through a process of rapid prototyping and fleet experimentation.

The Sea Trial process will develop enhanced warfighting capabilities for the fleet by more effectively integrating the thousands of talented and energetic experts, military and civilian, who serve throughout our Navy. Working together, we will fulfill the promise of "Sea Power 21."

Sea Trial Impact

- Fleet-led, enduring process of innovation
- Accelerated concept and technology development
- Enhanced headquarters/fleet alignment

Sea Warrior Impact

- Continual professional growth and development
- Improved selection and classification
- Interactive, web-based, incentivized detailing
- Networked, high-impact training

Sea Enterprise Impact

- Greater process efficiencies
- Divestment of non-core functions
- Organizational streamlining
- Enhanced investment in warfighting capability

Sea Warrior: Investing in Sailors

The Sea Warrior program implements our Navy's commitment to the growth and development of our people. It will serve as the foundation of warfighting effectiveness by ensuring the right skills are in the right place at the right time. Led by the Chief of Naval Personnel and Commander, Naval Education and Training Command, Sea Warrior will develop naval professionals who are highly skilled, powerfully motivated, and optimally employed for mission success.

Traditionally, our ships have relied on large crews to accomplish their missions. Today, our all-volunteer service is developing new combat capabilities and platforms that feature dramatic advancements in technology and reductions in crew size. The crews of modern warships are streamlined teams of operational, engineering, and information technology experts who collectively operate some of the most complex systems in the world. As optimal manning policies and new platforms reduce crew size further, we will increasingly need Sailors who are highly educated and expertly trained.

Introducing our people to a life-long continuum of learning is key to achieving our vision. In July 2001, we established Task Force EXCEL (Excellence through our Commitment to Education and Learning) to begin a revolution in training that complements the revolution in technologies, systems, and platforms for tomorrow's fleet. We are dedicated to improving our Sailors' professional and personal development, leadership, military education, and performance. Task Force EXCEL will apply information-age methods to accelerate learning and improve proficiency, including advanced trainers and simulators, tailored skills training programs, improved mentoring techniques, and more effective performance measurement and counseling tools. This growth and development focus will revolutionize the way we train.

Another initiative central to Sea Warrior is Project SAIL (Sailor Advocacy through Interactive Leadership). Project SAIL is moving the Navy toward an interactive and incentivized distribution system that includes guaranteed schools for high-performing non-rated personnel, team detailing, Internet job listings, an information call center, and expanded detailer outreach. These actions will put choice in the process for both gaining commands and Sailors, and it will empower our people to make more informed career decisions.

Our goal is to create a Navy in which all Sailors—active and reserve, afloat and ashore—are optimally assessed, trained, and assigned so that they can contribute their fullest to mission accomplishment.

Sea Enterprise: Resourcing Tomorrow's Fleet

Among the critical challenges that we face today are finding and allocating resources to recapitalize the Navy. We must replace Cold War-era systems with significantly more capable sensors, networks, weapons, and platforms if we are to increase our ability to deter and defeat enemies.

Sea Enterprise, led by the Vice Chief of Naval Operations, is key to this effort. Involving the Navy Headquarters, the Systems Commands, and the Fleet, it seeks to improve organizational alignment, refine requirements, and reinvest savings to buy the platforms and systems needed to transform our Navy. Drawing on lessons from the business revolution,

Sea Enterprise will reduce overhead, streamline processes, substitute technology for manpower, and create incentives for positive change. Legacy systems and platforms no longer integral to mission accomplishment will be retired, and we will make our Navy's business processes more efficient to achieve enhanced warfighting effectiveness in the most cost-effective manner.

Our Navy values operational excellence as its highest priority, and the vast majority of our training is devoted to sharpening tactical skills. However, it is also important that our leaders understand sound business practices so that we can provide the greatest return on the taxpayer's investment. To meet this need, we are creating educational opportunities to teach our leaders about executive business management, finance, and information technology. For example, the Center for Executive Education at the Naval Postgraduate School brings together rising flag officers and private industry decision-makers to discuss emerging business practices. We must also extend this understanding to the deckplates, so that our future leaders gain experience in a culture of strengthened productivity and continually measured effectiveness.

Increased inter-service integration also holds great promise for achieving efficiencies. For example, the Navy and Marine Corps tactical aviation integration plan will save billions of dollars for both services, enhance our interoperability, and more fully integrate our people. Whether it is the U.S. Coast Guard's Deepwater Integrated Systems Program, new munitions being developed with the U.S. Air Force, joint experiments with the U.S. Army on high-speed vessels, or a new combined intelligence structure with the U.S. Marine Corps, we will share technologies and systems whenever possible. Such efforts must not just continue; they must expand. Savings captured by Sea Enterprise will play a critical role in the Navy's transformation into a 21st-century force that delivers what truly matters: increased combat capability.

Our Way Ahead

The 21st century is clearly characterized by dangerous uncertainty and conflict. In this unpredictable environment, military forces will be required to defeat a growing range of conventional and asymmetric threats.

"Sea Power 21" is our vision to align, organize, integrate, and transform our Navy to meet the challenges that lie ahead. It requires us to continually and aggressively reach. It is global in scope, fully joint in execution, and dedicated to transformation. It reinforces and expands concepts being pursued by the other services-—ong-range strike; global intelligence, surveillance, and reconnaissance; expeditionary maneuver warfare; and light, agile ground forces—to generate maximum combat power from the joint team.

"Sea Power 21" will employ current capabilities in new ways, introduce innovative capabilities as quickly as possible, and achieve unprecedented maritime power. Decisive warfighting capabilities from the sea will be built around:

- Sea Strike—expanded power projection that employs networked sensors, combat systems, and warriors to amplify the offensive impact of sea-based forces;
- Sea Shield—global defensive assurance produced by extended homeland defense, sustained access to littorals, and the projection of defensive power deep overland;
- Sea Basing—enhanced operational independence and support for joint forces provided by networked, mobile, and secure sovereign platforms operating in the maritime domain.

The powerful warfighting capabilities of "Sea Power 21" will ensure our joint force dominates the unified battlespace of the 21st century, strengthening America's ability to assure friends, deter adversaries, and triumph over enemies—anywhere, anytime.

14

The Global War on Terrorism: The First 100 Days

"WE ARE SUPPORTED BY THE COLLECTIVE WILL OF THE WORLD."

President George W. Bush

The Coalition Information Centers
Washington, U.S.A.
London, U.K.
Islamabad, Pakistan

Table of Contents

Executive Summary

"The attack took place on American soil, but it was an attack on the heart and soul of the civilized world. And the world has come together to fight a new and different war, the first, and we hope the only one, of the 21st century. A war against all those who seek to export terror, and a war against those governments that support or shelter them."

—President George W. Bush, 10/11/01

On September 11, terrorists attacked freedom.

The world has responded with an unprecedented coalition against international terrorism. In the first 100 days of the war, President George W. Bush increased America's homeland security and built a worldwide coalition that:

- Began to destroy al-Qaeda's grip on Afghanistan by driving the Taliban from power.
- Disrupted al-Qaeda's global operations and terrorist financing networks.

- Destroyed al-Qaeda terrorist training camps.
- Helped the innocent people of Afghanistan recover from the Taliban's reign of terror.
- Helped Afghans put aside long-standing differences to form a new interim government that represents all Afghans - including women.

President Bush is implementing a comprehensive and visionary foreign policy against international terrorism. The President's policy puts the world on notice that any nation that harbors or supports terrorism will be regarded as a hostile regime.

Diplomacy. President Bush has built a worldwide coalition against terrorism. More than 80 countries suffered losses on September 11; 136 countries have offered a diverse range of military assistance; 46 multilateral organizations have declared their support; and with U.S. leadership and international support, Afghans are putting aside long-standing ethnic and political differences to form a new and representative government.

Terrorist Finances. The President fired the first shot in the war on terrorism with the stroke of his pen to seize terrorist financial assets and disrupt their fundraising pipelines. The world financial community is moving to starve the terrorists of their financial support. 196 countries support the financial war on terror; 142 countries have acted to freeze terrorist assets; in the U.S. alone, the assets of 153 known terrorists, terrorist organizations, and terrorist financial centers have been frozen; and major terrorist financial networks have been closed down.

The Military Campaign. Operation *Enduring Freedom* began on October 7, 2001, and enjoys the support of countries from the United Kingdom to Australia to Japan. The Taliban have been forced to surrender major cities. The military has destroyed 11 terrorist training camps and 39 Taliban command and control sites. And al-Qaeda terrorists have been captured, killed or are on the run.

Law Enforcement. The U.S. has led a global dragnet to help bring terrorists to justice and help prevent future terrorist acts, creating the Foreign Terrorist Tracking Task Force to prevent terrorists from entering the U.S.; arresting and indicting known terrorists; increasing the global sharing of law enforcement information; and implementing tough new anti-terrorism laws.

Humanitarian. As Afghanistan's largest humanitarian donor, the U.S. has increased its aid to the Afghan people by providing $187 million in aid since October alone, including food, shelter, blankets, and medical supplies. The President also launched the America's Fund for Afghan Children that has already raised more than $1.5 million for the children of Afghanistan. As the harsh Afghan winter approaches, the U.S. commitment to the Afghan people is saving lives.

Homeland Security. President Bush has taken steps to help protect America against further terrorist attacks, providing $20 billion for homeland security; strengthening intelligence efforts; creating the Office of Homeland Security and the Homeland Security Council; implementing tough new airline security measures; and taking steps to protect America's mail.

Helping the Survivors of September 11. The American people have responded with overwhelming compassion for the families of the victims of September 11, donating at least $1.3 billion to charities.

Respecting Islam. Almost immediately after the attacks the President took steps to protect Muslim-Americans from hate crimes. The President also held a series of events, including hosting the first-ever White House Iftar and an Eid event at the end of Ramadan; the President visited the Islamic Center; and the President created the "Friendship Through Education" initiative to bring American and Muslim children closer together.

The Tragedy of September 11

"Every one of the victims who died on September 11th was the most important person on earth to somebody."

—President George W. Bush, 12/11/01

On September 11 the terrorists committed an act of war against the innocent. The terrorists killed not only to end lives—they killed to end our way of life. Recently the terrorists said that we should forget the attacks of September 11. The terrorists would like nothing more than to silence the world's vocal opposition to their frightening vision they hope to export to every corner of the world.

The world will never forget the innocent victims, and the brave heroes who died attempting to save them. The world will never forget the survivors, the devastated families and the grieving friends they left behind:

- More than 3,000 people died or remain missing following the attacks. They came from more than 80 different nations, from many different races and religions.
- 343 firefighters and paramedics perished at the World Trade Center.
- 23 police officers and 37 Port Authority police officers died at the World Trade Center.
- Approximately 2,000 children lost a parent on September 11, including 146 children who lost a parent in the Pentagon attacks.
- One business alone lost more than 700 employees, leaving at least 50 pregnant widows.

On December 11, more than 120 countries stood together to remember the three-month anniversary of the terrorist attacks.

THESE NATIONS & AREAS SUFFERED LOSSES FROM THE SEPTEMBER 11 ATTACKS

Antigua & Barbuda	Ethiopia	Kenya	St. Kitts & Nevis
Argentina	France	Lebanon	St. Lucia
Australia	The Gambia	Liberia	St. Vincent & the
Austria	Germany	Lithuania	Grenadines
Bangladesh	Ghana	Malaysia	Sweden
Barbados	Greece	Mexico	Switzerland
Belarus	Grenada	The Netherlands	Taiwan
Belgium	Guatemala	New Zealand	Thailand
Belize	Guyana	Nicaragua	Togo
Bolivia	Haiti	Nigeria	Trinidad & Tobago
Brazil	Honduras	Pakistan	Turkey
Canada	Hong Kong	Panama	Ukraine
Chile	India	Paraguay	United Kingdom
China	Indonesia	Peru	United States of
Colombia	Ireland	Philippines	America
Czech Republic	Israel	Poland	Uruguay
Dominica	Italy	Portugal	Uzbekistan
Dominican Republic	Jamaica	Romania	Venezuela
Ecuador	Japan	Russia	Yemen
Egypt	Jordan	South Africa	Yugoslavia
El Salvador	Kazakhstan	South Korea	Zimbabwe
		Spain	
		Sri Lanka	

The Al-Qaeda Vision for the World

"... we calculated in advance the number of casualties from the enemy, who would be killed based on the position of the tower. We calculated that the floors that would be hit would be three or four floors. I was the most optimistic of them all ... due to my experience in this field, I was thinking that the fire from the gas in the plane would melt the iron structure of the building and collapse the area where the plane hit and all the floors above it only. This is all that we had hoped for."

—bin Laden

"This new enemy seeks to destroy our freedom and impose its views. We value life; the terrorists ruthlessly destroy it. We value education; the terrorists do not believe women should be educated or should have health care, or should leave their homes. We value the right to speak our minds; for the terrorists, free expression can be grounds for execution. We respect people of all faiths and welcome the free practice of religion; our enemy wants to dictate how to think and how to worship even to their fellow Muslims."

—President George W. Bush, 11/8/01

Al-Qaeda is a movement defined by hatred. They hate progress, and freedom, and choice, and culture, and music, and laughter, and women, and Christians, and Jews, and all Muslims who reject their distorted doctrines. They love and worship only one thing, and that is power—power they use without mercy to kill the innocent.

In Afghanistan, we have seen al-Qaeda's vision for the world. The leadership of al-Qaeda had great influence in Afghanistan and was supported by the Taliban regime. Afghanistan's people have been brutalized —many are starving and many have fled. Women were not allowed to attend school. A person could be jailed for owning a television. Religion could be practiced only as their leaders dictated. A man could be jailed in Afghanistan if his beard was not long enough.

The al-Qaeda terrorists believe it is acceptable to steal food meant for starving, innocent families. The al-Qaeda philosophy says it is acceptable to use innocent people as human shields for their military operations. The al-Qaeda philosophy says it is acceptable to oppress women and doom them to a lifetime of poverty.

- **Treatment of Women & Children.** First Lady Laura Bush led a worldwide initiative to highlight the Taliban's oppression of women. Before the Taliban, women played a key role in society. Then came al-Qaeda and their destruction of the Afghan family. The al-Qaeda-controlled Taliban regime especially targeted Afghan women and children, taking away their basic freedoms, splintering their families, putting their lives at risk, and relegating them to poverty. For example, the Taliban forbade the schooling for girls over the age of eight; shut down the women's university; banned women from working (stripping a society in desperate need of trained professionals of half its assets); restricted access to medical care for women; brutally enforced a restrictive dress code; forbade women from moving about their communities freely; and beat women for laughing out loud. The First Lady led a worldwide initiative to highlight the Taliban's oppression of women which helped lead to representation of women in the new interim government.

- **Targeting Civilians.** Al Qaeda and the Taliban regime have targeted civilians by literally using them as human shields for their military activities. For example, the November 6 *Washington Post* reports that the Taliban actually placed military assets in mosques and across the street from hospitals and innocent people's homes. Taliban commanders have also hijacked humanitarian aid facilities for military purposes. A senior officer told the *Washington Post*, "Whole villages are being used as human shields by the Taliban to protect their large stockpiles of ammunition and weapons hidden in nearby caves."

- **Humanitarian Crimes.** The al-Qaeda and Taliban contribution to the starving Afghan people has been a deliberate and systematic campaign to disrupt the efforts of international relief agencies to deliver desperately needed food and medical supplies to the Afghan people. For example, the Taliban seized control of two U.N. World Food Program (WFP) warehouses, one in Kabul, and one in Kandahar, containing more than half the World Food Program's wheat supply for Afghanistan. The WFP in Kandahar had been feeding 150,000 Afghans a month before the Taliban seizure. The Taliban are also actually hijacking humanitarian convoys for military purposes. The November 6 *Washington Post* reports, "A truck in a convoy purportedly on a humanitarian mission to deliver food tipped over, and crates of tank and mortar shells could be seen spilling to the ground underneath a thin layer of flour."

- **Al Qaeda & the Drug Trade.** Osama bin Laden and his organization finance many of their terrorist activities through the drug trade.

In fact, on October 25, 2001, *The Herald* (Glasgow) reported, "Osama bin Laden financed the development of a highly-addictive liquid heroin which he named 'The tears of Allah' as part of his multi-pronged terrorist campaign to destabilise western society. . . One source said yesterday: 'It should be called the Devil's Brew rather than Allah's tears. It is a one-way ticket to addiction and death.'" The United Nations has also weighed in on the Taliban and al Qaeda connection to the drug trade. According to a U.N. Committee of Experts report on Resolution 1333 (May 2001), "Funds raised from the production and trading of opium and heroin are used by the Taliban to buy arms and other war materiel, and to finance the training of terrorists and support the operations of extremists in neighbouring countries and beyond."

Diplomacy

"The message to every country is, there will be a campaign against terrorist activity, a worldwide campaign. And there is an outpouring of support for such a campaign. Freedom-loving people understand that terrorism knows no borders, that terrorists will strike in order to bring fear, to try to change the behavior of countries that love liberty. And we will not let them do that."

—President George W. Bush, 9/19/01

Since September 11, President Bush and Secretary of State Colin Powell have built a worldwide coalition for the war against terrorism. The coalition is stronger than ever and continues to grow.

- Since September 11, President Bush has met with leaders from at least 51 different countries to help build support for the war against terrorism.
- 136 countries have offered a range of military assistance.
- The U.S. has received 46 multilateral declarations of support from organizations.
- The U.N. General Assembly and Security Council condemned the attacks on September 12.
- NATO, OAS and ANZUS (Australia, New Zealand and the U.S.) quickly invoked their treaty obligations to support the United States.

Our NATO allies are assisting directly in the defense of American territory.

- 142 countries have issued orders freezing the assets of suspected terrorists and organizations.
- 89 countries have granted over-flight authority for U.S. military aircraft.
- 76 countries have granted landing rights for U.S. military aircraft.
- 23 countries have agreed to host U.S. forces involved in offensive operations.
- Through intelligence cooperation with many nations, we are acquiring evidence against those responsible for the attacks of September 11 and we are better able to prevent future attacks.
- With U.S. leadership and with international support, Afghans have put aside long-standing ethnic and political differences to form a new interim government, naming a president and 29 ministers with portfolio. The new government will also include women, who have been oppressed by the Taliban regime.
- On December 11, more than 120 nations around the world answered President Bush's call to reject terrorism and commemorate the victims of the September 11 attacks by holding remembrance ceremonies.
- The United States and several other allies have reopened embassies in Kabul.
- The President was joined by U.N. Secretary General Kofi Annan on November 11 for a memorial service honoring the citizens of all the countries killed in the World Trade Center.

Terrorist Finances

"We put the world's financial institutions on notice: if you do business with terrorists, if you support them or sponsor them, you will not do business with the United States of America."

—President George W. Bush, 11/7/01

Terrorists need money to carry out their evil deeds. The President's first strike in the war against terror was not with a gun or a missile—the President's first strike was with his pen as he took action to freeze terrorist finances and disrupt their pipelines for raising and moving money in the future.

The world's financial institutions have been put on notice—if you support, sponsor, or do business with terrorists, you will not do business with the United States. Denying terrorists access to funds is a very real success in the war on terrorism. Since September 11, the United States and its allies in the war on terrorism have been winning the war on the financial front:

- President Bush launched the first offensive in the war on terrorism on September 23 by signing an Executive Order freezing the U.S.-based assets of those individuals and organizations involved with terrorism.
- 196 countries and jurisdictions have expressed their support for the financial war on terror.
- 142 countries have issued orders freezing terrorist assets, and others have requested U.S. help in improving their legal and regulatory systems so they can more effectively block terrorist funds.
- The assets of at least 153 known terrorists, terrorist organizations, and terrorist financial centers have now been frozen in the U.S. financial system.
- Since September 11, the U.S. has blocked more than $33 million in assets of terrorist organizations. Other nations have also blocked another $33 million.
- On November 7, the U.S. and its allies closed down operations of two major financial networks—al-Barakaat and al-Taqwa—both of which were used by al-Qaeda and Osama Bin Laden as sources of income and mechanisms to transfer funds.
- On December 4, President Bush froze the assets of a U.S.-based foundation—The Holy Land Foundation for Relief and Development —that has been funneling money to the terrorist organization Hamas.
- The U.S. government created three new organizations—the Foreign Terrorist Asset Tracking Center (FTAT), Operation Green Quest and the Terrorist Financing Task Force. These new organizations will help facilitate information sharing between intelligence and law enforcement agencies and encourage other countries to identify, disrupt, and defeat terrorist financing networks.
- International organizations are key partners in the war on financial terrorism. On September 28, the United Nations Security Council passed resolution 1373 that requires all nations to keep their financial systems free of terrorist funds.
- The Financial Action Task Force—a 29-nation group promoting policies to combat money laundering—adopted strict new standards to deny terrorist access to the world financial system.

- The G-20 and IMF member countries have agreed to make public the list of terrorists whose assets are subject to freezing, and the amount of assets frozen.

The Military Campaign

"I said to the Taliban, turn them over, destroy the camps, free people you're unjustly holding. I said, you've got time to do it. But they didn't listen. They didn't respond, and now they're paying a price. They are learning that anyone who strikes America will hear from our military, and they're not going to like what they hear. In choosing their enemy, the evildoers and those who harbor them have chosen their fate."

—President George W. Bush, 10/17/01

Operation *Enduring Freedom*, the military phase, began October 7, 2001. Since then, coalition forces have liberated the Afghan people from the repressive and violent Taliban regime. As President Bush and Secretary of Defense Donald Rumsfeld have said, this is a different kind of war against a different kind of enemy. The enemy is not a nation —the enemy is terrorist networks that threaten the way of life of all peaceful people.

The war against terrorism is the first war of the 21st Century—and it requires a 21st Century military strategy. Secretary Rumsfeld has worked with our coalition allies and the courageous men and women of the U.S. military to craft a cutting-edge military strategy that minimizes civilian casualties, partners with local forces, and brings destruction to the oppressive Taliban who supported the al-Qaeda terrorist network.

The coalition has achieved broad military success while putting fewer than 3,000 U.S. ground troops on the ground in Afghanistan. And Secretary Rumsfeld and the U.S. military have also shown a lightning quick ability to adapt to a distant, harsh and ever-changing battlefield. In some cases, U.S. troops are conquering terrorists by welding together 21st Century technology with 19th Century tactics. Troops have chased terrorists on horseback while using mobile phones and global positioning systems to pinpoint targets for the Air Force. Bombers today use 21st Century targeting technology, and laser-guided and GPS guided smart bombs to destroy specific targets, including centuries—old caves used as terrorist headquarters.

While we've achieved a great deal of military success, much dangerous and difficult work remains to be done before the war on terrorism is won.

A few key military successes thus far in the war on terrorism include:

- In just weeks the military essentially destroyed al-Qaeda's grip on Afghanistan by driving the Taliban from power.
- Taliban leaders have surrendered major cities to opposition forces, including Kandahar, Kabul, Kunduz, and Mazar-e-Sharif.
- The military has destroyed at least 11 terrorist training camps and 39 Taliban command and control sites. The Wall Street Journal reported on December 13 that as many as 50,000 terrorists from more than 50 countries may have received training in al-Qaeda camps in Afghanistan in recent years.
- About 2.5 million humanitarian rations have been dropped to aid the people of Afghanistan.
- U.S. Marines have established a military base at Kandahar airport.
- Routes are being blocked to try to prevent the escape of al-Qaeda and Taliban members.
- Senior al-Qaeda and Taliban officials have either been captured or killed.
- The U.S. military rescued two American Christian aid workers who were being held as prisoners by the Taliban.
- Friendship Bridge between Afghanistan and Uzbekistan was reopened to transport humanitarian aid by land.
- Minefields and roads are being cleared to ensure delivery of aid and freedom of movement.
- Leaflet drops and radio broadcasts continue daily to convey our determination, provide truthful information, and encourage the capture of Osama bin Laden.

The military action in Afghanistan represents a global coalition effort. In addition to the United States, military assets are being deployed from many other nations, including the United Kingdom, Australia, Canada, Czech Republic, France, Germany, Italy, Japan, New Zealand, Poland, Russia and Turkey.

Law Enforcement

"Terrorists try to operate in the shadows. They try to hide. But we're going to shine the light of justice on them. We list their names, we publicize their pictures, we rob them of their secrecy. Terrorism has a face, and today we expose it for the world to see."

—President George W. Bush, 10/10/01

The U.S. is leading a global dragnet to help bring terrorists to justice and help prevent future terrorist acts.

Prevention and Investigation:

- As of December 17, 460 individuals were being detained by the INS. 116 individuals, 77 of whom are in custody, are facing federal criminal charges —including Zacarias Moussaoui who has been charged with conspiring with Osama bin Laden and al-Qaeda to murder thousands of innocent people in New York, Virginia and Pennsylvania.
- The Department of Justice (DOJ) created the new 22 "Most Wanted Terrorists" list.
- The FBI created a national task force to centralize control and information sharing resulting in hundreds of thousands of leads, over 500 searches, thousands of interviews of witnesses, and numerous court-authorized surveillance orders.
- The U.S. government has offered a reward of up to $25 million for information leading directly to the apprehension or conviction of Osama bin Laden.
- The Treasury Department and the Department of Justice collaborated to freeze the assets and accounts of 62 individuals and organizations connected with two terrorist-supporting financial networks, the al-Taqua and the al-Barakaat, and one organization funneling money to Hamas.
- The Department of State strengthened its "Rewards for Justice Program" which authorizes the Secretary of State to offers rewards of more than $5 million for information that prevents acts of international terrorism against the United States. The State Department has also launched a series of Public Service Announcements to educate the American public on the program.
- Improved information sharing between the law-enforcement and intelligence communities, allowing nationwide search warrants for e-mail and subpoenas for payment information, and to place those who access the Internet through cable companies on the same footing as everyone else.
- At the Attorney General's request, the State Department designated 39 entities as terrorist organizations.
- The U.S. has forged new cooperative agreements with Canada to pro-

tect our common borders and the economic prosperity they sustain.

- Created 93 Anti-Terrorism Task Forces—one in each U.S. Attorney's district—to integrate the communications and activities of local, state and federal law enforcement.
- Created the Foreign Terrorist Tracking Task Force to focus on preventing terrorists from entering the country, and to locate and remove those who already have.
- The Department of Justice crafted a new reorganization plan— Reorganization and Mobilization of the Nation's Justice and Law Enforcement Resources—which is DOJ's strategy for fiscal years 2001 to 2006 to help meet the new anti-terrorism mission.
- Reorganization of the Immigration and Naturalization Service (INS) to reform the agency's structure by separating its service and enforcement functions. Fulfills President Bush's pledge to improve the efficiency and effectiveness of the nation's immigration system.
- The Department of Justice launched the Responsible Cooperators Program. Justice will provide immigration benefits to non-citizens who furnish information to help apprehend terrorists or to stop terrorist attacks.
- INS arrested Mazen Al Najjar after he was ordered to be deported for violating his visa, had established ties to terrorist organizations and held leadership positions in the Islamic Concern Project and the World and Islam Studies Enterprise.
- Zayd Hassan Abd Al-Latif Masud Al Safarini was arrested for his indictment in 1991 for the September 5, 1986, hijacking of Pan American World Airways Flight 73, demonstrating DOJ's commitment to track down terrorists no matter how long it takes.

Civil Rights:

- The President moved swiftly to protect Muslims from hate crimes and the Department of Justice followed his lead by having their Civil Rights Division sponsor community forums in Chicago, Illinois, and Dearborn, Michigan, on combating ethnically motivated violence as a result of the September 11th terrorist attacks.
- Attorney General John Ashcroft and AAG for Civil Rights Ralph F. Boyd, Jr. have met with 29 prominent leaders from the Arab and Muslim American and Sikh communities and underscored DOJ's strong commitment to investigate and prosecute violators of federal hate crime laws.

- The Department of Justice, the U.S. Equal Opportunity Commission and the Department of Labor issued a joint statement against employment discrimination in the aftermath of September 11.
- Investigating approximately 300 incidents involving violence, or threats of violence against individuals perceived to be of Middle Eastern origin. Federal charges have been brought in 6 cases, coordinating with local prosecutors in at least ten instances where cases are being prosecuted locally.

Victim Relief:

- Provided approximately $52 million in assistance to victims and their families and $10 million in emergency assistance to the NYPD.

Humanitarian Relief

"Ultimately, one of the best weapons, one of the truest weapons that we have against terrorism is to show the world the true strength of character and kindness of the American people. Americans are united in this fight against terrorism. We're also united in our concern for the innocent people of Afghanistan."

President George W. Bush, 10/11/01

The humanitarian situation in Afghanistan remains dire. Millions face the threat of starvation. 70% of the Afghan people and 1/2 of all Afghan children are malnourished. Only 13% of the Afghan people have access to clean water.

Years of civil war—compounded by the rule of the Taliban and the worst drought in 30 years—have made matters worse. The Taliban were clearly more interested in protecting al-Qaeda than feeding the starving, innocent people of Afghanistan. Al-Qaeda and the Taliban have not only failed to provide security, food, and shelter for the Afghan people, but they have also disrupted the efforts of international relief agencies to deliver desperately needed food and medical supplies to the Afghan people. Among other things, the Taliban have seized and looted humanitarian supplies for themselves, and have harassed and beaten Afghan and international aid workers.

The typically harsh Afghan winter is arriving and the U.S., with its international partners, is doing everything it can to help bring hope to the innocent Afghans who have suffered under the brutal and oppressive al-Qaeda and Taliban regime:

- Even before September 11, the U.S. was the leading humanitarian aid donor for Afghanistan.
- Last fiscal year the U.S. provided $183 million of humanitarian assistance alone to Afghanistan.
- Since the beginning of October alone, the U.S. has provided more than $187 million in humanitarian assistance to Afghanistan.
- On October 10, USAID Administrator Natsios announced a five-point assistance strategy for Afghanistan: reduce death rates; minimize population movements; lower and then stabilize food prices; ensure that aid reaches those it is intended for; and begin developmental relief programs.
- As of mid-December the international community, led by the U.S., has delivered 127,368 metric tons (MT) of food aid to Afghanistan, using, trucks, boats, barges, aircraft, and thousands of people to overcome numerous logistical and security obstacles. (For context, 52,000 MT of food will feed approximately six million people for one month.)
- Between October 7 and December 13, the Department of Defense airdropped 2,423,700 Humanitarian Daily Rations (HDRs) to Afghans who could not be reached by relief workers because of ongoing conflict. The entire operation cost approximately $51 million.
- The President announced the creation of the America's Fund for Afghan Children. The President asked American children to send $1 dollar—or whatever they can afford—to the Fund to help buy important humanitarian supplies. America's children have donated more than $1.5 million thus far to the Fund. The first shipment of humanitarian goods purchased from this fund left the U.S. for the Afghan children on Sunday, December 9.
- The government has provided more than $62 million in grants to support relief activities in Afghanistan. The programs include supporting agriculture, rehabilitating water resources, funding health services, repairing shelters, and providing critical non-food items such as blankets, tents and kitchen sets. Additional grants have funded UN coordination efforts and a radio program to provide humanitarian and security information to Afghans in their home languages.

- USAID has provided funding for wool blankets and quilts, shelter kits, plastic sheeting and winterized tents. Further, USAID is distributing mattresses, clothes, stoves, cooking sets, firewood, coal, lanterns and water containers.
- The government has provided medical kits and funds for health centers and mobile clinics in Afghanistan and is sponsoring public heath education and programs on hygiene, obstetrics, maternal and childcare, and malnutrition. USAID is employing trained personnel to conduct educational outreach on basic health and nutrition, especially to women. USAID is helping expectant mothers, training local birth attendants and funding the distribution of vitamins and the immunization of young children.
- The government has provided funds for rehabilitation and reconstruction in the areas of housing, roads and bridges, wells and irrigation systems, agriculture and food security, and initiating "food for work" and "food for cash" initiatives.
- USAID has funded six airlifts of critical commodities to Afghanistan. The airlifts have provided shelter materials, tents, health supplies and high-energy food items for vulnerable people in Afghanistan.
- The State Department has provided $32,260,000 to relief agencies to assist Afghan refugees in Pakistan, Iran and other neighboring countries. The grants also provide funds to assist refugees attempting to return to their homes in Afghanistan.
- The government has sent Disaster Assistance Response Team (DART) personnel to Pakistan, Uzbekistan, Turkmenistan and Tajikistan to ensure that relief efforts are effective and well coordinated.

Respecting Islam

"The Islam that we know is a faith devoted to the worship of one God, as revealed through The Holy Qu'ran. It teaches the value and importance of charity, mercy, and peace."

—President George W. Bush, 11/15/01

The United States is a nation of religious freedom, and the President has acted to ensure that the world's Muslims—from Dearborn, Michigan to Kabul, Afghanistan—know that America appreciates and celebrates the rich traditions of Islam:

- At the national prayer service following the September 11 attacks, the President included religious leaders from many faiths, including an Imam from the Islamic Society of North America. Subsequently the President hosted an interfaith meeting on September 20 with leaders of different religious denominations to pray jointly for the victims of the September 11 tragedies and called for national reconciliation.
- Soon after the terrorist attacks, the President visited the Islamic Center of Washington to meet with American Muslim leaders and deliver a message of tolerance and solidarity. The President condemned unwarranted attacks on Americans of Muslim faith, and urged Americans to show their support for their Muslim friends.
- President Bush launched the "Friendship Through Education" initiative, encouraging children in America and children in Muslim nations to connect through email, letter writing, and different friendship and understanding projects. The President wants this initiative to help youths from different societies deepen their understanding of each others' traditions and outlooks.
- For Ramadan, on November 19 President Bush hosted the first-ever Iftar—or breaking-of-the-fast—dinner at the White House, which included the ambassadors from nations with Muslim populations. The President also issued a warm greeting to Muslims around America and around the world with a special Ramadan message.
- The State Department asked U.S. embassies in Muslim countries to host Iftar dinners and many members of his administration also held their own celebrations. Secretary of State Colin Powell, Attorney General John Ashcroft, and Deputy Secretary of Defense Paul Wolfowitz all participated in Iftar dinners.
- On December 17, President Bush hosted Muslim children at the White House in honor of Eid al-Fitr, Islam's most sacred holiday. He read an Eid book to the children and hosted them for cookies and punch as well as delivering a present to each child in the tradition of Eid. The President also issued a taped Eid message and read an Eid greeting.
- Secretary of Energy Spencer Abraham recognized Americans for acts of compassion following the September 11th attacks—he honored, for example, a church that started an escort service for Muslim women who wear the hijab, and a citizen who created a fund to assist

low-income Muslim victims of hate-inspired vandalism. Secretary Abraham will also launch a series of public service announcements further promoting tolerance.

Homeland Security

"We face a united, determined enemy. America is going to be prepared."

—President George W. Bush, 10/8/01

President Bush has taken action to help protect America against terrorist attacks. The government is working around the clock to protect Americans. Among many other steps:

- The President worked with Congress to provide $20 billion to promote homeland security, including funds to upgrade intelligence and security, provide recovery assistance to disaster sites, help victims' families, increase numbers of law enforcement personnel, provide health care for displaced Americans, and purchase irradiation equipment to sanitize the mail.
- The President established the Office of Homeland Security—under the diligent guidance of Governor Tom Ridge—and the Homeland Security Council to coordinate, and implement the Executive Branch's efforts to detect, prevent, protect against, respond to, and recover from terrorist attacks within the United States.
- The President moved to implement tough new airline security standards that tighten background checks for airline screeners and workers, dramatically expands the federal air marshal program, creates strict new baggage security requirements, and tightens security in all areas of airports.
- The President established an advisory committee for cyber security to ensure that America's key infrastructures are protected. The advisory committee is a public/private partnership.
- The Administration has strengthened coordination between law enforcement agencies of the U.S. and neighboring countries to address common threats while ensuring the free flow of goods and people.
- The Food and Drug Administration has enhanced the food screening process of imported foods.

- The Department of Health and Human Services created the Office of Public Health Preparedness, to coordinate the national response to public health emergencies.
- Public health professionals provided antibiotics to more than 30,000 people to protect against their possible exposure to anthrax.
- HHS increased the supply of drug caches around the country, added specific use drugs, and began to increase the supply of small pox vaccine to 300 million.
- The President created a Presidential Task force to help Americans prepare in their homes, neighborhoods, schools, and other public places from the consequences of terrorist attacks.
- The Centers for Disease Control and the U.S. Postal Service provided guidelines on how to handle mail that had potentially come into contact with anthrax.
- EPA has worked with water utilities, chemical, pesticide, petroleum and fertilizer manufacturers to increase their vigilance and secure their resources against an attack.

The Survivors of September 11

"It is said that adversity introduces us to ourselves. This is true of a nation as well. In this trial, we have been reminded, and the world has seen, that our fellow Americans are generous and kind, resourceful and brave. We see our national character in rescuers working past exhaustion; in long lines of blood donors; in thousands of citizens who have asked to work and serve in any way possible."

—President's Remarks at National Day
of Prayer and Remembrance, 9/14/01

Every one of the victims who died on September 11th was the most important person on earth to somebody. The American people have responded to the tragedies of September 11 with an unprecedented outpouring of support for their fellow Americans who lost so much on September 11:

- While it is impossible to catalog every penny of contributions, at least $1.3 billion has been collected in aid for families of both civilian and uniformed victims of the September 11th terrorist attacks.
- There are many relief organizations collecting aid to distribute to the

families of the thousands of victims from the September 11th terrorist attacks. Examples include the Twin Towers Orphan fund, the Pentagon Assistance Fund, the WTC School Fund, the Washington Redskins Relief Fund, the Dole-Clinton Families of Freedom Scholarship Fund, and funds established by the United Jewish Communities, Catholic Charities, the Burn Center at the Washington Hospital Center, and many, many more.

- The American Red Cross has raised over $647.4 million and has distributed over $217.7 million to the families of the victims of September 11th.

- The September 11th United Way Relief Fund has been working thoughtfully and deliberately to distribute $143 million in cash and services to help rebuild the lives of victims' families and affected communities.

- On September 1, the entertainment industry came together in historic fashion to raise funds and raise the spirits of all who have been touched by the horrific tragedy that has struck America. Over $150 million was pledged through the United Way September 11th Telethon Fund, "America: Tribute to Heroes." By the end of the year, a total of $100 million in cash assistance will have been delivered to victims families through the Telethon Fund by the entertainment industry.

- New York City's major human service organizations have joined forces under an umbrella group called the 9/11 United Services Group. At the urging of the New York State Attorney General, the group launched a confidential database, which will serve as a central clearinghouse for information on victims, survivors, financial needs, and the amounts of money and services provided by charities. This database enables the different charities to communicate with each other and track both the needs of the families of victims as well as the amount of assistance they are receiving.

- More than 100 people with ties to the Sept. 11 terror attacks will carry the Olympic torch as it makes its way to Salt Lake City for the Winter Games. The torch will be passed at the Pentagon on Friday, December 21, 2001.

- The President announced www.libertyunites.org in the Rose Garden on September 18, 2001, and praised Americans for their outpouring of charitable relief support in the wake of September 11th.

15

Organizational Structure

(Reprinted from the May 2005 Naval Review issue of *Proceedings,* courtesy of the U.S. Naval Institute)

Organizational

U.S. National Defense Command Structure

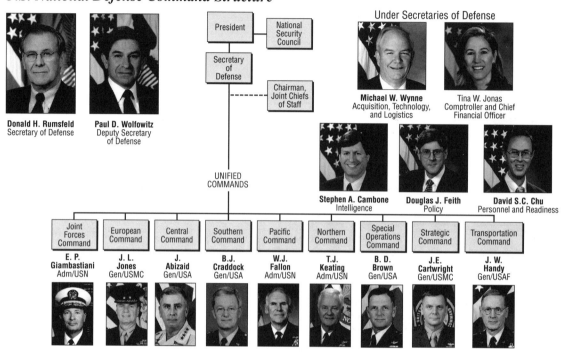

Under Secretaries of Defense

Donald H. Rumsfeld
Secretary of Defense

Paul D. Wolfowitz
Deputy Secretary
of Defense

President — National Security Council

Secretary of Defense

Chairman, Joint Chiefs of Staff

Michael W. Wynne
Acquisition, Technology,
and Logistics

Tina W. Jonas
Comptroller and Chief
Financial Officer

Stephen A. Cambone
Intelligence

Douglas J. Feith
Policy

David S.C. Chu
Personnel and Readiness

UNIFIED COMMANDS

Joint Forces Command	European Command	Central Command	Southern Command	Pacific Command	Northern Command	Special Operations Command	Strategic Command	Transportation Command
E. P. Giambastiani Adm/USN	**J. L. Jones** Gen/USMC	**J. Abizaid** Gen/USA	**B.J. Craddock** Gen/USA	**W.J. Fallon** Adm/USN	**T.J. Keating** Adm/USN	**B. D. Brown** Gen/USA	**J.E. Cartwright** Gen/USMC	**J. W. Handy** Gen/USAF

Office of the Secretary of Defense

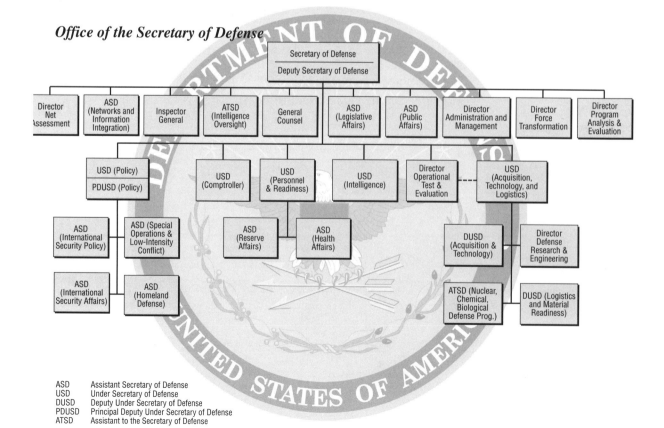

Secretary of Defense

Deputy Secretary of Defense

Director Net Assessment

ASD (Networks and Information Integration)

Inspector General

ATSD (Intelligence Oversight)

General Counsel

ASD (Legislative Affairs)

ASD (Public Affairs)

Director Administration and Management

Director Force Transformation

Director Program Analysis & Evaluation

USD (Policy)
PDUSD (Policy)

USD (Comptroller)

USD (Personnel & Readiness)

USD (Intelligence)

Director Operational Test & Evaluation

USD (Acquisition, Technology, and Logistics)

ASD (International Security Policy)

ASD (Special Operations & Low-Intensity Conflict)

ASD (Reserve Affairs)

ASD (Health Affairs)

DUSD (Acquisition & Technology)

Director Defense Research & Engineering

ASD (International Security Affairs)

ASD (Homeland Defense)

ATSD (Nuclear, Chemical, Biological Defense Prog.)

DUSD (Logistics and Material Readiness)

ASD   Assistant Secretary of Defense
USD   Under Secretary of Defense
DUSD   Deputy Under Secretary of Defense
PDUSD Principal Deputy Under Secretary of Defense
ATSD   Assistant to the Secretary of Defense

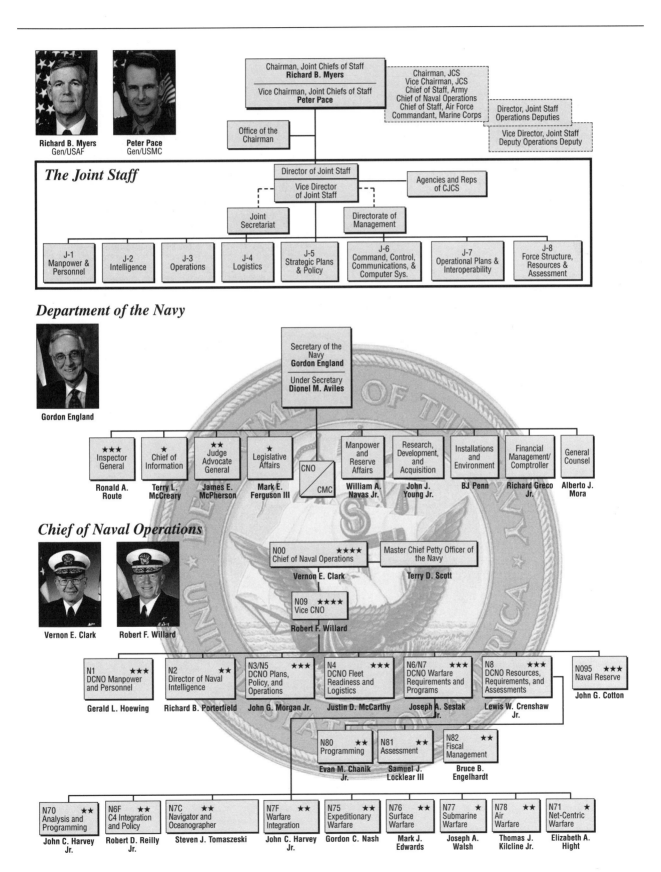

Richard B. Myers
Gen/USAF

Peter Pace
Gen/USMC

Chairman, Joint Chiefs of Staff
Richard B. Myers

Vice Chairman, Joint Chiefs of Staff
Peter Pace

Office of the Chairman

Chairman, JCS
Vice Chairman, JCS
Chief of Staff, Army
Chief of Naval Operations
Chief of Staff, Air Force
Commandant, Marine Corps

Director, Joint Staff
Operations Deputies

Vice Director, Joint Staff
Deputy Operations Deputy

The Joint Staff

Director of Joint Staff

Vice Director
of Joint Staff

Agencies and Reps
of CJCS

Joint
Secretariat

Directorate of
Management

J-1 Manpower & Personnel	J-2 Intelligence	J-3 Operations	J-4 Logistics	J-5 Strategic Plans & Policy	J-6 Command, Control, Communications, & Computer Sys.	J-7 Operational Plans & Interoperability	J-8 Force Structure, Resources & Assessment

Department of the Navy

Gordon England

Secretary of the Navy
Gordon England

Under Secretary
Dionel M. Aviles

★★★ Inspector General	★ Chief of Information	★★ Judge Advocate General	★ Legislative Affairs	CNO / CMC	Manpower and Reserve Affairs	Research, Development, and Acquisition	Installations and Environment	Financial Management/ Comptroller	General Counsel
Ronald A. Route	**Terry L. McCreary**	**James E. McPherson**	**Mark E. Ferguson III**		**William A. Navas Jr.**	**John J. Young Jr.**	**BJ Penn**	**Richard Greco Jr.**	**Alberto J. Mora**

Chief of Naval Operations

Vernon E. Clark

Robert F. Willard

N00   ★★★★
Chief of Naval Operations

Vernon E. Clark

Master Chief Petty Officer of
the Navy

Terry D. Scott

N09   ★★★★
Vice CNO

Robert F. Willard

N1 ★★★ DCNO Manpower and Personnel	N2 ★★ Director of Naval Intelligence	N3/N5 ★★★ DCNO Plans, Policy, and Operations	N4 ★★★ DCNO Fleet Readiness and Logistics	N6/N7 ★★★ DCNO Warfare Requirements and Programs	N8 ★★★ DCNO Resources, Requirements, and Assessments	N095 ★★★ Naval Reserve
Gerald L. Hoewing	**Richard B. Porterfield**	**John G. Morgan Jr.**	**Justin D. McCarthy**	**Joseph A. Sestak Jr.**	**Lewis W. Crenshaw Jr.**	**John G. Cotton**

N80 ★★ Programming	N81 ★★ Assessment	N82 ★★ Fiscal Management
Evan M. Chanik Jr.	**Samuel J. Locklear III**	**Bruce B. Engelhardt**

N70 ★★ Analysis and Programming	N6F ★★ C4 Integration and Policy	N7C ★★ Navigator and Oceanographer	N7F ★★ Warfare Integration	N75 ★★ Expeditionary Warfare	N76 ★★ Surface Warfare	N77 ★ Submarine Warfare	N78 ★★ Air Warfare	N71 ★ Net-Centric Warfare
John C. Harvey Jr.	**Robert D. Reilly Jr.**	**Steven J. Tomaszeski**	**John C. Harvey Jr.**	**Gordon C. Nash**	**Mark J. Edwards**	**Joseph A. Walsh**	**Thomas J. Kilcline Jr.**	**Elizabeth A. Hight**

Organizational

Headquarters, U.S. Marine Corps

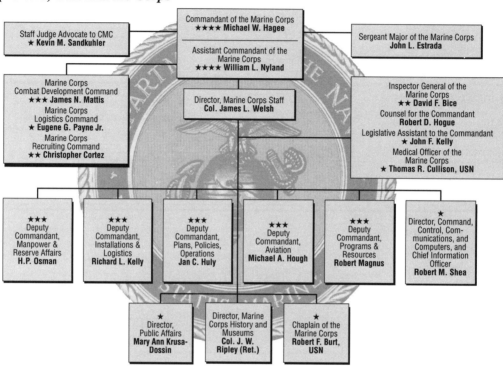

Commandant of the Marine Corps
★★★★ **Michael W. Hagee**

Assistant Commandant of the Marine Corps
★★★★ **William L. Nyland**

Staff Judge Advocate to CMC
★ **Kevin M. Sandkuhler**

Sergeant Major of the Marine Corps
John L. Estrada

Marine Corps
Combat Development Command
★★★ **James N. Mattis**
Marine Corps
Logistics Command
★ **Eugene G. Payne Jr.**
Marine Corps
Recruiting Command
★★ **Christopher Cortez**

Director, Marine Corps Staff
Col. James L. Welsh

Inspector General of the
Marine Corps
★★ **David F. Bice**
Counsel for the Commandant
Robert D. Hogue
Legislative Assistant to the Commandant
★ **John F. Kelly**
Medical Officer of the
Marine Corps
★ **Thomas R. Cullison, USN**

★★★
Deputy
Commandant,
Manpower &
Reserve Affairs
H.P. Osman

★★★
Deputy
Commandant,
Installations &
Logistics
Richard L. Kelly

★★★
Deputy
Commandant,
Plans, Policies,
Operations
Jan C. Huly

★★★
Deputy
Commandant,
Aviation
Michael A. Hough

★★★
Deputy
Commandant,
Programs &
Resources
Robert Magnus

★
Director, Command,
Control, Com-
munications, and
Computers, and
Chief Information
Officer
Robert M. Shea

★
Director,
Public Affairs
**Mary Ann Krusa-
Dossin**

Director, Marine
Corps History and
Museums
**Col. J. W.
Ripley (Ret.)**

★
Chaplain of the
Marine Corps
**Robert F. Burt,
USN**

U.S. Coast Guard

As of 1 June 2005

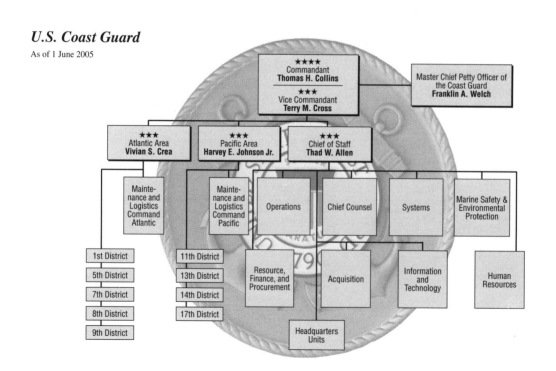

★★★★
Commandant
Thomas H. Collins
★★★
Vice Commandant
Terry M. Cross

Master Chief Petty Officer of
the Coast Guard
Franklin A. Welch

★★★
Atlantic Area
Vivian S. Crea

★★★
Pacific Area
Harvey E. Johnson Jr.

★★★
Chief of Staff
Thad W. Allen

Mainte-
nance and
Logistics
Command
Atlantic

Mainte-
nance and
Logistics
Command
Pacific

Operations

Chief Counsel

Systems

Marine Safety &
Environmental
Protection

1st District
5th District
7th District
8th District
9th District

11th District
13th District
14th District
17th District

Resource,
Finance, and
Procurement

Acquisition

Information
and
Technology

Human
Resources

Headquarters
Units